Handbook of Organic Food Processing and Production

Second Edition

Edited by

Simon Wright

and

Diane McCrea

**Blackwell
Science**

Copyright © 2000 Blackwell Science Ltd
Editorial Offices:
Osney Mead, Oxford OX2 0EL
25 John Street, London WC1N 2BS
23 Ainslie Place, Edinburgh EH3 6AJ
350 Main Street, Malden
    MA 02148 5018, USA
54 University Street, Carlton
    Victoria 3053, Australia
10, rue Casimir Delavigne
    75006 Paris, France

Other Editorial Offices:

Blackwell Wissenschafts-Verlag GmbH
Kurfürstendamm 57
10707 Berlin, Germany

Blackwell Science KK
MG Kodenmacho Building
7–10 Kodenmacho Nihombashi
Chuo-ku, Tokyo 104, Japan

First edition published 1994 by Chapman & Hall
Second edition published 2000

Set in 10/12 pt Times
by Sparks Computer Solutions Ltd, Oxford
http://www.sparks.co.uk
Printed and bound in Great Britain by
MPG Books Ltd, Bodmin, Cornwall

DISTRIBUTORS

Marston Book Services Ltd
PO Box 269
Abingdon
Oxon OX14 4YN
(*Orders:* Tel: 01235 465500
           Fax: 01235 465555)

USA
Blackwell Science, Inc.
Commerce Place
350 Main Street
Malden, MA 02148 5018
(*Orders:* Tel: 800 759 6102
                781 388 8250
           Fax: 781 388 8255)

Canada
Login Brothers Book Company
324 Saulteaux Crescent
Winnipeg, Manitoba R3J 3T2
(*Orders:* Tel: 204 837-2987
           Fax: 204 837-3116)

Australia
Blackwell Science Pty Ltd
54 University Street
Carlton, Victoria 3053
(*Orders:* Tel: 03 9347 0300
           Fax: 03 9347 5001)

A catalogue record for this title
is available from the British Library

ISBN 0-632-05541-3

Library of Congress
Cataloging-in-Publication Data
is available

For further information on
Blackwell Science, visit our website:
www.blackwell-science.com

List of contributors

Francis Blake, Standards and Technical Director, Soil Association, Bristol House, 40–56 Victoria Street, Bristol BS1 6BY.

Peter Challands, Marketing Director, Deans Foods, Bridgeway House, Upper Ickneild Way, Tring, Hertfordshire HP23 4JX.

David Crucefix, Ekopro Partnership, Ctra. Les Planes km4, nave 4, 17150 Sant Gregori, Girona, Spain.

John Dalby, Certification Executive, Organic Farmers and Growers Ltd, c/o 50 High Street, Soham, Ely, Cambridgeshire CB7 5HP.

Robert Duxbury, Technical Product Manager – Organic Foods, Sainsbury's Supermarkets Ltd, Stamford House, Stamford Street, London SE1 9LH.

Renee Elliott, Founder, Planet Organic, 42 Westbourne Grove, London W2 5SH.

Carolyn Foster, Institute of Rural Affairs, University of Wales, Aberystwyth, Llanbadarn Campus, Aberystwyth, SY23 3AL.

Louise Hemsted, CROPP/Organic Valley, 507 W. Main Street, La Farge, WI 54639, USA.

Andrew Jedwell, Founder, Meridian Foods, Hoel Gauad, Cynwyd, Denbighshire LL21 0NE.

Bob Kennard, Graig Farm Organics, Dolau, Llandrindod Wells, Powys LD1 5TL.

Scott Kinear, Organic Federation of Australia Inc., c/o 452 Lygon St., East Brunswick, Victoria 3057, Australia.

David Lanning, General Manager – Agriculture, Lloyd Maunder Ltd, Willand, Cullompton, Devon, EX15 2PJ.

Richard Maunder, Director, Lloyd Maunder Ltd, Willand, Cullompton, Devon, EX15 2PJ.

Diane McCrea, Consultant in Food and Consumer Affairs, 17 Vernon Road, London N8 0QD.

Michael Michaud, Market Gardener, Sea Spring Farm, Lyme View, West Bexington, Dorcehester, Dorset DT2 9DD.

Mark Redman, Barrow Hill Farm Cottage, Barrow Hill, Stalbridge, Sturminster, Newton, Dorset DT10 2QX.

Craig Sams, President, Whole Earth Ltd, 2 Valentine Place, London SE1 8QH.

Professor Sir Colin Spedding, Chairman 1988–99, UKROFS, Vine Cottage Hurst, Reading, Berkshire RG10 0SD.

Andrew Whitley, Founder of The Village Bakery, Melmerby, Penrith, Cumbria CA10 1HE.

Simon Wright, The Organic Consultancy, 101 Elsenham Street, London SW18 5NY.

Contents

Foreword

The first edition of this handbook was published in 1994, predating the current organic boom by several years. Sainsbury's was selling organic food before this. It all started in 1986 when a Sainsbury's technologist called Robert Duxbury visited a small vegetable farmer in West Wales called Patrick Holden to discuss the possibility of Sainsbury's buying Patrick's organic vegetables. Today Robert is our Technical Product Manager for Organic Foods for Primary Agriculture and Patrick is the Director of the Soil Association, the UK's foremost organic certification body.

Both Robert and the Soil Association have contributed chapters to the second edition of this most useful handbook, which is designed to guide the many processors and manufacturers who would like to produce organic food and drink to be sold in stores around the world such as Sainsbury's.

The world of organic food processing and food production is complex and highly regulated, and I congratulate the authors of this book for the attention to detail they have shown in their various chapters. The UK is currently the most dynamic market for organic food and drink in the world, so it is appropriate that many chapters are written by UK producers. The global perspective required for successful organic production is reflected in the chapters written by authors from America and Australia.

A theme that runs through this book is the necessity of partnerships between the different participants in organic production – between grower and processor, and between processor and retailer. Here Sainsbury's has led the way. In 1997 we set up our Organic Resourcing Club (SOuRCe), an innovative group of key organic suppliers to Sainsbury's who meet regularly in order to develop sales of organic food sold through our stores. In 1999 Sainsbury's agreed a deal with a group of UK organic dairy farmers to take all the milk they could produce for a period of five years at a guaranteed price. This type of partnership deal demonstrates our commitment to the continued growth of the organic sector. We have many more organic projects underway – for an update please visit the Organic section of our website at www.tasteforlife.co.uk/organics.

Whether you are new to organic food production or an experienced organic processor I am confident you will find this book of great value and I recommend it to anyone wanting to know more about this fascinating and fast-moving sector.

Ian Merton
Director of Fresh & Convenience Foods
Sainsbury's Supermarkets

1 Introduction

Craig Sams

1.1 Introduction

At the threshold of the twenty-first century, the historically unprecedented wealth that has been created on the planet has brought a material standard of living for many that was undreamed of in any previous era. It has also brought environmental and ecological problems that dwarf those which led to the fall of ancient civilisations in the once Fertile Crescent of the Middle East. Our technology and engineering skills have enabled us to make habitable those parts of the planet that were hitherto wilderness, but we have also seen the Dust Bowl in the American midwest, the drying out of the Aral Sea, the desertification of African savannah and the disappearance of tropical rainforest. As the world's population expands the pressure on resources increases and the finite nature of those resources becomes more apparent. There are no new worlds to conquer and diminishing areas of forest to turn into new pasture or arable land. The question of how humankind can sustain its place on earth takes on a new urgency. In the discussion on sustainability, organic farming plays a pivotal role. The continuing conquest of nature requires ever more precarious advances, as the pesticides and hybrid seeds of the 'green revolution' fail to fulfil expectations and hopes turn to genetic engineering and synthesised foods as the way forward.

1.2 Organic farming

What's in a name? To those of a scientific bent, 'organic' describes that branch of chemistry where carbon atoms are present in a molecule's structure. To the farmer, however, the term describes a way of growing food crops.

To many people 'organic' just means the old-fashioned way of farming. However, most of the world's deserts are the product of old-fashioned farming methods. If humankind is a parasite on Earth, then like every successful parasite, we have a vested interest in maintaining a sustainable balance with our host. All organic farming is based on this key assumption of sustainability. If one takes something out of the soil, one must return it. If there is life in earth, then that life must be understood, respected and supported. The nutrients and humus that are removed from the soil by a plant should be replaced with an equal or greater value of nutrients and organic matter. Anyone can clear land, grow crops on it until the land is exhausted and eroded, and move on, but only when available land is infinite. Agriculture based on non-sustainable use of natural resources can be just as harmful to the environment as the use of artificial fertilisers and pesticides. It is not what we define as organic farming, even though it may be free of chemical inputs.

Organic farming assumes that the soil is a living entity and that success must come from nurturing and encouraging its life.

'Organic living' is the ultimate outcome of the organic philosophy. The principles of sustainability, returning to the earth what is taken away, and recycling apply to all aspects of production and consumption beyond the field and garden. Organic agriculture becomes the foundation of a philosophy that seeks a sustainable future for life on Earth.

At the root of an understanding of organic farming is an understanding of the soil. The structure of the soil determines what can be grown and how fertility can be maintained. The structure of the soil can be enhanced to increase the efficiency whereby nutrients are created, retained, and taken up by plants. Only when the quantity and fertility of the soil are maintained or increased can farming be truly called organic.

At one extreme is desert, soil in which there is little humus, little moisture, and little invertebrate or vertebrate life. At the other extreme is humus-rich, moisture-retentive soil, teeming with life including fungi and bacteria, and rich in decaying vegetable matter. In such soil plants can not only put down a strong root structure to draw up nutrients, but the fungi that surround the root hairs can proliferate and play their role in synthesising nutrients to feed the plant above. There is a complex of relationships that is so intricate that terms like 'magic' are sometimes called upon to describe it. To some it is a sacred process, the whole transcending its parts because of some indefinable element that brings it all together. No wonder that at the root of all religions, ancient and modern, is a wonder at the miracle of fertility and a healthy respect for the processes that confer it or take it away. Religious festivals celebrate the processes of germination, growth and transmutation. The power of rain, sun and earth are venerated as part of the intellectual and spiritual process that seeks to understand and to harness their awesome energies.

Conventional farming does not eschew the basics of organic farming. Soil structure and dynamics, fertility, drainage, and rotation – all are considered. But when there is an economically viable short cut that frees the farmer from the need to work in harmony with the laws of nature, the conventional farmer will take it. The conventional farmer does not consider the macrocosm, does not take into account the impact on the outside world of actions on the microcosm of the farm.

Conventional farming is dependent upon, and addicted to, outside inputs. The mechanism of addiction is universal: when an external input provides something to a system that is normally derived from within the system, then that internal source tends to atrophy. In conventional farming the farmer has an inevitable tendency to become addicted to the use of artificial external inputs. The more chemical fertilisers are used, the less that the fertility-building processes within the soil can function. Weedkiller is needed to suppress weeds which flourish on fertiliser inputs, and these damage the balance of life in the soil. Abundant foliage growth encourages fungal disease, leading to a need for fungicides. The more often a crop is grown in the same soil year after year, protected by pesticides, the less viable that soil becomes to support any other plant life.

What does it matter, as long as the crops keep coming in? The 1993 floods in Iowa, USA, uncovered fossils from the Devonian era. Soil that was formed at a time when the inhabitants of Earth were still mostly fishes has been eroded away forever as a result of agricultural practices that commenced in the 1870s. When soil disappears into the sea it cannot be replaced. Chemical fertilisers are only effective when there is some topsoil which they can enhance. Once that topsoil is gone, there are rapidly diminishing returns and the cost of producing foods on barren sands and clays fertilised with artificial fertilisers becomes uneconomic or impossible. Extremes dominate as flooding and drought alternate.

During the floods of 1993, American organic farmers could feel smug. Their land had a knitted texture that protected it from erosion. When the waters subsided, they did not experience the same problems of 'panning' (where the soil forms a hard crust that does not allow water to drain through), as their conventional neighbours. Their soil acted as a sponge, absorbing water but not losing precious humus and fertility. Their farming methods had passed the ultimate challenge of sustainability. In 1999 severe drought in the eastern United States led to substantial crop losses among soybean farmers. Yet the Rodale Institute Farming Systems Trial reported yields from their trial organic plots of 30 bushels/acre compared to 16 bushels/acre from their conventional 'high-yield' plots and credited the moisture-retaining quality of organic soil structure for the difference.

An organic farmer builds fertility by incorporating animal manures, vegetable humus, and natural rock and vegetable extracts including seaweed manures into the soil. Fertility is increased by green manuring where the foliage of green crops is ploughed into the soil. Ploughing in the roots and stalks of harvested crops further enhances the humus structure and water-absorbing capacity of the soil. When crops are rotated on a 4- or 5-year cycle, insects and fungal or bacterial diseases do not have the opportunity to become entrenched. Weeds are outwitted by planting seeds at night, or by letting the weeds germinate and grow before ploughing them in. In all of it, there is an acceptance that sometimes the insects or the weeds will reduce the ultimate yield.

The difference between conventional farming and organic farming is the difference between war and peace. Conventional farmers wage war on nature, winning from her what they can, using their armoury of chemicals to keep her at bay while they take as much as they can get. Organic farmers attempt to apply a creative process of conflict resolution whereby nature volunteers her bounty in return for a balancing contribution towards her well-being. E.M. Shumacher wrote: 'We speak of the battle with Nature, but we would do well to remember that if we win that battle, we are on the losing side' (Shumacher 1973).

1.3   Conversion

When a conventional farmer decides to farm organically, it is not an overnight conversion. A minimum and somewhat arbitrary period in the UK of two years has been established as the time it takes to convert to an organic way of farming. It takes time to build a healthy soil structure, to rid the soil of toxic pesticide residues, and to develop natural fertility. During the conversion period the farmer will experience reduced yields and will need to put much of what he grows back into the ground in order to create a balanced healthy soil. The cost of conversion depends upon how severely the fertility and the health of the land has been diminished by conventional practices. Once the virtuous cycle of replenishment that characterises organic agriculture is under way, the farmer begins to be able to farm profitably using organic methods. The output from the farm is healthy, free of pesticide residues, and the farm has joined the ranks of the fully organic. Of course, the farmer may find that the range of crops that can be grown is limited by soil and environmental conditions that might previously have been overcome with the appropriate chemicals.

1.4   Biodynamic farming

The Austrian philosopher Rudolf Steiner (1861–1925) believed that humans today are capable of rediscovering a higher level of spiritual awareness and thereby become able to participate in the spiritual processes of the world in the way that our ancestors did in prehistory. His philosophy, anthroposophy, goes beyond religion, where the world of spirit is touched only via the medium of priests, and into a world of direct spiritual experience. In agriculture and horticulture he developed the biodynamic method, which harnesses unseen forces and works in harmony with lunar and solar periods to seek to attain better food quality and good yields. Biodynamic farming matches the criteria of organic farming and goes further, with a record of successful results.

1.5   The cost of organic farming

Until the period following World War II, few farmers depended to any great extent upon chemical fertilisers. At a pinch, fertilisers might help amend careless or shortsighted practices, or restore fertility where flooding or overgrazing may have depleted it, but in general farmers found it more economic to use their own resources and operated a mixed farming system that precluded the need to spend hard-earned money on the products of the chemical and pharmaceutical industries. They would optimise, rather than maximise, production. What has changed to create such an ascendancy for conventional farming?

From the mid-1940s onwards, governments in most developed countries began to engage themselves more deeply in the lives of their citizens. Housing, employment, education and agriculture were just a few areas of human endeavour where the individual found the hand of government reaching more deeply into daily affairs. With the stated goal of increasing productivity and of providing economic security for those who owned and farmed the land, the government became increasingly involved in training, technology, marketing and distribution within the agricultural sector. This resulted in a sense of responsibility for the sector's well-being and led to price fixing, tariffs on imports, and subsidies. Guaranteed prices came to dominate the sector to an unprecedented extent. As market forces became less relevant, producers no longer faced the insecurity of not knowing what price they might get for their output. This led inevitably to overproduction, and a commercial advantage to the farmer who used increased amounts of agricultural chemicals to achieve it.

Lucrative grants to cover the cost of draining wetlands, grubbing out hedgerows and ploughing downland pasture were offered and eagerly taken, and low-grade land was drafted into arable production. Imports of high-quality bread wheats and low-quality feed grains from abroad fell as they were replaced by domestic production. The cost to the British economy was significant. The diversion of capital from the manufacturing and service sectors to the agricultural sector contributed to industrial decline, falling exports and growing unemployment. Britain's capital was being locked away in the value of land. When Britain entered the European Community (EC) and became part of the Common Agricultural Policy (CAP), there was a massive wave of investment by financial institutions in land. The CAP provided a virtual guarantee that land values would rise, and they did. The industrial and service sectors of the economy paid the price.

1.6   How subsidies make organic farming uncompetitive

Agricultural chemicals have been around since the mid-nineteenth century, when Liebig developed artificial forms of nitrogen. Yet the use of fertilisers and pesticides was not widespread until the period following World War II. The world then had significant overcapacity for the production of nitrate fertilisers, using manufacturing facilities that had produced nitroglycerine explosives in wartime. In addition, poisons like DDT and phosphene were in abundant supply after wartime investment in the means of their production. However, the financial incentives did not exist to warrant a wholesale switch to intensive chemical agriculture and farmers preferred to save the cash outlays on input costs and instead maintain fertility by natural methods.

During the wartime era, people's awareness of their dependency on imported food was heightened. The long lines of supply from the old colonies were seen as risky, in the event of another war. The 'Dig for Victory' campaign had created a psychological bias towards national self-sufficiency in food. When government policy sought to achieve and maintain self-sufficiency in food, people thought it was a 'good thing' and failed to consider the cost, or alternative approaches to achieve the same end. A political system of agricultural support was developed which rewarded production and favoured the use of chemical inputs. It is the legacy of this system that applies today.

While the exact figures vary from crop to crop, the basic rules are the same. A farmer's profit per acre is the difference between input costs and gross revenues. When the cost of land is higher, yields or prices (or both) need to be higher to compensate. When Britain joined the CAP, with higher guaranteed prices for farm output, land prices quickly rose to reflect the increased profitability of farming. Hedgerow removal gathered pace and the 'prairification' of East Anglia reached its peak. Investment funds, including pension funds and property groups, switched their assets into agricultural property.

1.7   How does it work?

When prices are kept artificially high, it becomes more economic to use chemical inputs to increase yields. Whether one farms organically or not, there is no difference in the capital costs of land and equipment, the cost of ploughing, cultivation, feeding the farmer's family and staff, and tending to animals. Chemical inputs are not cost-effective unless the income from the extra production gained is greater than the cost of those chemical inputs. By maintaining artificially high prices, Western governments have ensured that there is a strong incentive to use chemical inputs. Of course, if prices fell, all farmers would be worse off. This has underpinned the political justification for price supports. There is a powerful argument for giving farmers financial support. Most Western societies enjoy a broad consensus that the countryside should be populated by people who grow food and tend the land. However, if farmers were encouraged to grow quality food, and to maintain the land in good condition, the cost would ultimately be less to the taxpayer, and the environmental, international and public health benefits would be a bonus.

A basic assumption is that the organic grower will harvest at worst two-thirds of the yield obtained by the conventional grower. The economics of using chemical inputs depend on prices high enough to justify the use of those chemicals. Table 1.1, using

Table 1.1 Effect of EC wheat subsidy on income from sale of milling wheat produced from conventional and organic farming.

	EC		World	
	Conventional	Organic	Conventional	Organic
Yield per acre (tonnes)	3	2	3	2
Price per tonne	£140	£140	£90	£90
Gross income	£420	£280	£270	£180
Chemicals	£100	0	£100	0
Rent	50	50	50	50
Seed	20	20	20	20
Net income per acre	£250	£210	£100	£110

wheat as an example, illustrates the way that price supports make the use of chemical inputs more profitable.

At EU prices, the non-organic farmer is at a £40 per acre disadvantage. The organic price premium seeks to restore the balance, and if the organic farmer receives an extra £20 per tonne for his output then he is on a par with his conventional counterpart. If there were no price supports, both farmers would be worse off, but the organic farmer would be the better off of the two, and the consumer and the environment would benefit.

At this point a social decision must be made. How do we, as a society, wish to support agriculture and the principle of self-sufficiency? One way is to abandon all price supports and pay the farmer who undertakes to farm organically an annual payment per acre farmed that compensates for the reduced income arising from non-use of chemical inputs. This payment allows the market price for agricultural produce to apply. The non-organic farmer would also require an acreage payment, but it could be much lower to reflect the external costs of non-organic farming and to encourage conversion to organic methods.

Supporting chemical-based agriculture has an additional cost in environmental degradation. It also diverts investment away from genuinely profitable enterprises and into the fertiliser and pesticide industries that exist solely because of the bias of agricultural policy. This kind of market distortion does not make a nation more competitive: it is comparable to military investment in that there is no real return on the investment other than the presumption of survival rather than extinction. This is the argument on which the post-Second World War involvement of government in agriculture in Europe and North America has been based – self-sufficiency.

The cost to society and the consumer of the European Union price subsidy is huge:

- The cost of food on the shelf is higher
- The environment is degraded
- Small farms disappear and are replaced by larger scale agribusiness
- Rural employment declines and social stability suffers
- The quality of food deteriorates
- The health risks of pesticides lead to an increase in degenerative diseases
- Access to the countryside is restricted and, at some seasons, dangerous

- Tax revenues are diverted into sustaining land values and the turnover of agricultural chemical manufacturers
- Production gains in arable crops mask economic damage to other industries, such as marine fisheries and honey production
- The fertility of land is exploited to produce surpluses that are not needed, thereby wasting a productive asset's future value
- The cost of removing excess nitrates and pesticides from the water supply increases the end-user price
- Large surpluses of end products of farming such as meat, butter, and alcohol made from grapes and grain are kept in long-term storage. Many of these surpluses are given away as 'aid' or burned as fuel in the form of ethanol or 'bio-diesel' from rapeseed oil. In the US alone, over 3m hectares of land are now wastefully devoted to growing corn for ethanol fuel production
- Of the CAP budget of €40 billion, administrative and fraud policing costs account for 25%, storage costs a further 25%, with the remaining 50% going to farmers, mainly used for the purchase of chemical inputs.

If farmers were paid to be custodians of the countryside, with a payment based on acreage farmed organically and with negative incentives for chemical use, supply could achieve an equilibrium with demand. Farmers would grow what was required for the market, and government would reward them for not polluting the landscape and groundwater supplies. The aesthetic and amenity value of the countryside would be enhanced.

1.8   First World agriculture and Third World poverty

The global ramifications of subsidised conventional agriculture are also significant. The value of agricultural land is ultimately a reflection of the value of what it will grow. The value of a food crop is a reflection of how much land it takes to grow a given amount of that crop. In simplistic terms, if an acre of land produces 2t of apples or 4t of plums, then the market price of plums will be half the market price of apples.

When the EU and the United States dump subsidised food on Third World countries they depress the market price for food and the value of land. The southward march of desert in the Sahara illustrates this. Marginal farmers in the Sahel who carefully nurture water supplies, build walls and terraces to prevent erosion, and protect cultivated land from grazing animals are driven out of business because the millet and sorghum they grow cannot compete in the market with wheat that is a 'gift' from the unwitting taxpayers of France or the USA. They abandon their land, which becomes grazing land for the flocks and herds of nomadic peoples before finally succumbing to the advance of the desert.

The export of subsidised cheap food affects other producers too. The value of all food crops in the world is depressed by the artificially low price of basic commodities such as soybeans, maize, and wheat. It becomes more economic for Third World producers to grow cash crops than to grow subsistence crops which can be bought in cheaply from the EU and the USA. This leads to the decline of subsistence farming, dispossession from the land, and the growth of plantation farming with waged labour replacing individual land ownership. Plantation agriculture is by definition monoculture, whether it be citrus, cocoa, coffee, or beef. It is dependent on pesticides to deal with the inevitable spread of insect pests and disease when the natural balance has been lost, and on fertilisers to achieve the higher yields necessary to compensate for the additional cost of paying waged labour. The value of plantation crops is also depressed: if soybeans and rapeseed are subsidised, then the value of oil palm is depressed to the artificially low level of vegetable oils. The value of all agricultural land and its produce in the unsubsidised Third World reflects the distorted economics of the heavily subsidised industrial economies. The value of all Third World produce is depressed, increasing the pressure on natural resources.

1.9   Fair trade and organic farming

When Rachel Carson's *Silent Spring* was published, Scandinavian governments led the way in banning persistent pesticides such as DDT. However, the use of such pesticides has actually increased since then, as chemical farming techniques are exported to the Third World. Under programmes of agricultural aid and technical assistance there has been a widespread adoption of chemical methods in Third World countries. Modern cacao growing exemplifies the change. Malaysia has converted large areas of rainforest and expired rubber plantations to cacao production, and in Brazil large plantations have replaced small family-run plots. Instead of growing alongside other trees and plants, large monoculture areas are planted with cocoa trees, which are then heavily fertilised and kept alive by regular spraying with fungicides. The plantation workers suffer pesticide-related diseases and the women labourers have a high rate of miscarriages and deformed births. On some plantations women have to produce proof of sterilisation before they are allowed to work. Child labour is often drafted in to help parents earn enough at piece rates. Similar conditions apply in tea, coffee, oil palm, and other tropical crop production. A reaction to this has been 'fair trade'. Often this may mean nothing more than conscience-salving actions like issuing protective clothing to plantation workers. It may also mean dealing only with co-operatives, attempting to bypass the large plantation companies, the trading intermediaries, and the government monopolies.

Like 'organic', 'fair trade' can mean anything to anyone, and the need has arisen for a definition that can guarantee to the consumer that the 'fair trade' product they purchase genuinely does not involve exploitation. The definition of fair trade overlaps with the definition of organic. To the Third World farmer the power of the multinationals goes hand in hand with the power of the agrichemical companies. Generous prices encourage farmers to borrow to invest in expanded production. When these prices fall, the farmers find that they have wasted natural fertility or are stuck with hybrid varieties of plants, so becoming more dependent on chemical inputs. To finance the higher level of input costs in chemical farming, farmers must borrow. Many find themselves unable to repay debt and are dispossessed. They join the landless agricultural workers, seeking employment on large plantations, or follow the exodus to the cities.

There is often not even a purely economic justification for this dislocation. Research at the University of the Philippines and the International Rice Research Institute has shown that the gain in extra rice production achieved by using high input chemical farming methods is outweighed solely by the additional healthcare costs arising from those methods (*New Scientist* 30 October 1993). Add the cost of imported inputs, environmental degradation and erosion and the balance tips even further.

In Holland, Germany and Britain, organisations like the Max Havelaar Foundation, Transfair, and the Fairtrade Foundation have established standards and provide inspection and certification services for fairly traded products. Although their criteria fall short of organic standards, they do prohibit the use of the most undesirable pesticides that are already banned in the EU. In addition they encourage direct trading relationships with grower co-operatives and marketing groups for family-run farms, while still allowing for plantation grown commodities where social criteria are met. The harmonisation of fair trade standards mirrors the development of organic standards, and a single fair trade symbol for use in all European countries is planned. IFOAM (International Federation of Organic Agricultural Movements) incorporate 'social justice' provisions in their standards for organic agriculture and processing, to require fundamental fair trade principles.

1.10   History of the organic movement in the UK

Deep in the human psyche there has always been a yearning for a bygone golden age, for a time when abundance was the norm and winning food was not a labour. By the time Cain was condemned to till the soil as punishment for slaying Abel, the ideal of a lost harmony with nature had already taken root. When William Cobbett rode out from Kensington to survey the countryside of South England, he returned to propose in *Cottage Economy* a way to protect and sustain the rural landscape by keeping agricultural units small, with each agricultural labourer owning his own land, or at least having security of tenure. This 'small is beautiful' approach has characterised much of what has followed and has often been condemned for being impractical or whimsical. However, the 'bigger is better' approach of chemical farming has its own failings. As these failings became more apparent, an organised concept of organic farming emerged. It was closely linked with the back-to-nature movement of the early part of the twentieth century.

In the early 1900s H.J. Massingham was a regular visitor to the Notting Hill home of W.H. Hudson, the field naturalist and ornithologist and founder of the Royal Society for the Protection of Birds. Hudson's popular novel, *Green Mansions*, foresaw the burning of the rainforests and the eradication of indigenous cultures that became one of the most shameful features of the century. Influenced by Hudson's mystical reverence for nature and his concern at the disappearing culture of rural England, Massingham helped found the Kinship of Nature, a movement whose adherents included Sir Albert Howard, Rolf Gardiner and Sir Robert McCarrison.

Sir Albert Howard spent 30 years in India, until 1931, and combined a scientific training with a study of traditional composting methods of India and China. In *An Agricultural Testament* (Howard 1940), he advocated that Britain preserve the 'cycle of life' and adopt 'permanent agriculture' systems, using urban food waste and sewage to build soil fertility and to produce nutritious food to build a healthy nation. *An Agricultural Testament* was reprinted several times, in American and British editions, and profoundly influenced thinking on both sides of the Atlantic. One person who was greatly affected by Howard's writings was J.I. Rodale, who coined the term 'organic' to describe this approach to agricultural production and went on in 1942 to publish *Organic Gardening and Farming*, America's most influential organic growing magazine. Howard urged that 'The first place in post-war plans of reconstruction must be given to soil fertility in every part of the world.'

His call inspired Lady Eve Balfour (Fig. 1.1) to undertake the Haughley experiment. At the age of 12, in 1910, she had already decided to become a farmer. By 1919, armed with an Agricultural Diploma from the University of Reading, she and her elder sister were farming at Haughley, in Suffolk. In 1938 she met Sir Albert Howard and was deeply influenced by his ideas. Her interest went beyond the health of the soil, incorporating nutritional ideas of healthy diet based on whole foods grown in healthy soil.

The debate at that time was still rooted in the nineteenth-century argument between the supporters of Baron Justis von Liebig, a chemist who believed that the mineral content of soil was the only factor governing fertility, and those who believed that humus-rich 'living soil' was the key to fertility and to healthy plants which would not succumb to erosion, disease and pests. To resolve the debate, Lady Balfour used her own farm at Haughley and a neighbouring farm donated by her neighbour, Alice Debenham, to perform comparative research between organic and non-organic farming methods (Balfour 1943).

As World War II drew to a conclusion, Lady Balfour set about establishing the Soil Association, a pioneering organic farming charity, in November 1946. The founding aims of the Soil Association were:

(1)   To bring together all those working for a fuller understanding of the vital relationships between plant, animal and man.

Fig. 1.1   Lady Eve Balfour.

(2) To initiate, co-ordinate and assist research in this field.
(3) To collect and distribute the knowledge gained so as to create a body of informed public opinion.

The farm at Haughley became a test bed for organic farming theories and an inspiration for farmers reluctant to follow the path signposted by the Ministry and the agrichemical companies. The Soil Association membership established Britain's first organic food shop, Wholefood, in 1960, which became an outlet for the organic produce they grew. The larger part of their organic output still had to be sold on the commercial market, but as the health food industry grew, more and more of their output was sold labelled as organic, so that consumers could choose for themselves. In the early 1960s the influence of *Silent Spring* triggered a greater awareness of the environmental damage that was the result of modern agricultural methods. At the same time, William Longood's *The Poisons in Your Food* first made consumers aware of the personal hazards involved in eating food that had been doused with chemicals from the field to the supermarket shelf.

By the late 1960s the environmental movement had taken root and a much larger and non-specialist audience had come to accept the principles underlying organic farming. The natural food stores that proliferated in Europe and the USA in the 1970s made organic produce the focal point of their selling proposition and a viable commercial market for organically grown food was born. A definition of 'organically grown' became necessary as opportunists jumped on the bandwagon, drawn by the price premiums that organic products could attract. In 1974 the Soil Association established the first set of organic standards. These standards formed the foundation of EU Regulation 2092/91, the first legally enforceable definition of the term 'organic'. The world now is following the same pattern, to ensure that in international trade the equivalent standard for 'organic' applies in whichever country a food is grown or produced. In the absence of a statutory standard in the USA, American farmers who wish to export to the EU or Japan comply with EU or IFOAM standards.

1.11 UK market development

In Britain, the retail market for organic food traces its roots to the Soil Association's Wholefood shop in London. The Wholefood shop had a butcher division, located around the corner from the main shop to protect vegetarian sensibilities. Lilian Schofield, the manager, maintained that the retail operation would not have been viable without the butcher shop.

The proliferation of natural foods stores in the early 1970s led to more rapid growth in the consumption of organically grown grains and pulses, and bakeries such as Ceres pioneered the production of organic bread. The swing to organic eating did not have a great impact on most British organic farmers. Few natural foods stores could cope with the problems of handling fresh produce, and almost all had ethical objections to trading in meat. The organic food they sold was mostly imported organically grown grains, seeds and pulses.

In the mid-1980s the supermarkets, responding to the 'greening' of their customers and to vociferous demands from pressure groups, dabbled tentatively in the market. Leading them was Safeway, where the commitment came from the top. Chief executive Alistair

Grant saw that if Safeway were identified with organic foods, the consumer would identify Safeway with organic quality. In practice sales were disappointing. Nonetheless key supermarkets persevered, often selling organic produce at a loss, but unwilling to lose the business of the high-calibre consumers attracted by organic produce. In 1992 one supermarket analysed its sales and found that organic foods occupied 1.5% of shelf space, yet only represented 0.5% of sales turnover. However, it also found, when it analysed the 'basket value' of total purchases of consumers, that those consumers who purchased one or more organic items also spent on average twice as much per store visit as those consumers who did not. In the battle with cheap food discounters specialising in 2000–3000 best-selling branded lines, an important competitive weapon for supermarkets is variety of choice. Organic foods, as part of a range of 15 000–20 000 lines, are a key element in persuading consumers to shop in their stores.

The supermarkets made their primary commitment to fresh produce, dairy products and meat, exploiting the key sectors which had been neglected by the natural foods stores but which were developing rapidly through box schemes. These are also the categories where consumer anxiety about conventional production methods is strongest. With scares about Alar on apples, lindane in carrots, BSE in cows, hormones in milk, and anxieties about animal welfare on intensive farms, consumers did not have to be 'deep green' to take the precaution of buying organic when the appropriate goods were made readily available.

From 1988 to 1993 sales of organic foods rose 320% from £21m to £105m. (see Table 1.2). Growth by the end of 1993 was slowing to only 10% per annum and a Mintel report conservatively predicted that the market would increase by a further 50% by 1998, to a projected level of £150m (Mintel 1993). Sales in 1998 exceeded £390m, a 271% increase in the 5-year period. By 1999 every major multiple had an organic offering in all or most of its stores.

The consumer commitment to organic food is lasting. Once a consumer has made the decision to purchase a particular item or category of product from organic sources, they find it difficult to revert to buying the non-organic equivalent. One supermarket's research describes a 'cycle of adoption' in which most consumers begin to purchase organic in the produce category, then move in succession to include dairy, meat, bread, grocery products and eventually clothing in their organic shopping (Ozminskyj 1999). Brand loyalty is a long-established and constant factor in marketing. Organic foods command a depth of loyalty that would be envied by branded manufacturers. Paradoxically, this loyalty occurs most strongly in those product areas such as meat, fresh produce and dairy produce, where brand loyalty in the conventional market is least strong. Together these sectors

Table 1.2   Value of UK organic food sales 1988, 1992, 1998 (£m).

	1988 (£m)	1992* (£m)	1998† (£m)	% increase 1998 over 1993
Produce	—	66	175	165
Dairy products	—	4	63	1500
Meat products	—	9	14	55
Grocery and processed	—	13.5	138	900
Total	21	92.5	390	320

*Mintel (1993); †Soil Association (1999)

comprise 63% of total sales of organic foods, further emphasising the contribution that supermarket involvement has made to the growth of the organic market. In the late 1980s much UK-grown organic produce ended up being sold into the conventional market with no organic identification; by the late 1990s more than 70% of sales in this category were imported as UK organic producers could not keep up with demand.

The impact of the EU Organic Food Regulation 2092/91 underpinned consumer confidence in the validity of organic claims and has provided the foundation on which the organic market could develop (see Chapter 2). When the consumer knows that the price premium reflects a legally defined and enforceable standard, then parting with the extra money is easier. The failure to achieve equivalence of standards in meat production may be a partial explanation for the substantially lower rate of growth in this category since 1993. Other factors may include a disproportionately high number of vegetarians converting to organic food as well as supply bottlenecks for some meat products and a confusion of perception between 'free range' and 'organic.'

In 1997 the second wave of supermarket commitment to the organic market gathered momentum. Management structures that had been developed to deal with big manufacturer suppliers were modified to introduce the flexibility needed to deal with the smaller but vitally important organic producers and processors. In a race to be first to capture the organic consumer, leading supermarkets expanded their organic offering and by 1998 had obtained a 62% share of the UK organic market.

To some extent the involvement of the supermarkets represents a lost opportunity for the natural foods retailers that emerged from the wholefood movement. In the United States, natural food retailers did not hesitate to include organic fresh produce including meat and dairy products in their offering. The additional sales volume and customer traffic that they enjoyed enabled expansion to supermarket-sized outlets, and large natural foods supermarkets in major American population centres can compete effectively with supermarkets. Consumers in the UK who want to eat organic can more easily obtain most of their requirements at a supermarket and then top up their purchases at specialist independent outlets. In the United States such consumers are more likely to be able to satisfy their shopping requirements without ever needing to visit a conventional supermarket. Conventional supermarkets in the US have been less successful in attracting customers who are seeking organic foods as many of these target consumers are obtaining their requirements from an organic full-scale supermarket. Internet retailers are also making inroads into the market, using fresh produce and regular delivery as the key to selling a much wider range. The role of the internet in the future development of retail organic food selling will see competition between the 'clicks and mortar' operators who can offer traditional shopping integrated with a web-based offering and the virtual retailers who will enjoy lower overheads if they can achieve sufficient sales volume.

1.12 Producers, processors and marketeers

Organic food and farming operates on many levels, and has pioneered some new ways of food production that are now becoming the norm. There are several levels of production at which value is added. Every stage must be controlled and monitored to ensure that organic standards are maintained from the farmer's field through to the final consumer.

- *Production.* This is the foundation level, planting a seed, nurturing it to the fulfilment of its destiny as a fruit, root, or more seeds, and harvesting it.
- *Post-harvest processing.* This involves grading, cleaning, and storing in hygienic conditions that ensure freedom from pests, moulds and other detrimental influences.
- *Packaging.* Fresh produce or whole grains are packed either on farm, or more likely, at a central packing station. The producer has now passed ownership of his product to a packer or onward processor. It is at this stage that regulations protect the integrity of the producer's product.
- *Processors.* Organic inspection and certification ensures that processors operate proper controls to ensure that the claims on packaging for using organic raw materials are accurate.
- *Contract processing.* One of the reasons that organically produced foods are more expensive has little to do with the real, on-farm costs of production. These can often be marginal, with the long-term cost of farming organically competitive with the cost of high-input agriculture. However, the small scale of handling, through to processing, amplifies the cost differentials instead of reducing them as would be the case with most foodstuffs. The small processor inevitably has higher overheads to spread over a smaller amount of production, and must recover this cost in the selling price. When demand for a particular organic processed product (tinned baked beans, for example) reaches a level where it is economic to produce them on the larger scale production runs required in large canning plants, then the price differential between organic baked beans and non-organic falls dramatically. This also applies to organic bread and other products where investment in efficient production technology brings cost savings that far outweigh the additional costs of organic raw materials.

Like the medieval tradesmen who banded together into guilds, the traders who have made the commitment to organic production often find it profitable to transcend their competitive differences and band together to advance their mutual interest. In the UK the Soil Association acts as an umbrella body for them and increasingly the business of promoting the market for organic products is seen as one that can best be dealt with on a generic level. The Soil Association has raised its public profile and the organic message is being delivered by well-informed media to a public whose understanding of the issues is increasingly sophisticated. As more and more consumers understand and support the rationale for eating organic, the market size increases. The decision as to which organic products to purchase then becomes a choice among the offerings of competing suppliers.

References

Balfour, Lady E. (1943) *The Living Soil.* Faber & Faber, London.
Howard, Sir A. (1940) *An Agricultural Testament.* Oxford University Press, Oxford.
Mintel (1993) *Vegetarians and Organic Food.* Mintel, London.
New Scientist (30 October 1993) (research by University of Philippines and International Rice Research Institute).
Ozminskyj, A. (1999) *Organic – as Natural as Nature Intended.* Tesco Strategic Planner, Cheshunt.
Pretty, J. (1999) *An Assessment of the External Costs of UK Agriculture.* University of Essex.

Shumacher, E.F. (1973) *Small is Beautiful: A Study of Economics as if People Mattered.* Blond & Briggs, London.

Soil Association (1999) *The Organic Food and Farming Report 1999.* Soil Association/Baby Organix, Bristol.

2 International Legislation and Importation

David Crucefix and Francis Blake

2.1 Introduction

At the start of the twenty-first century it is hard to deny the increasing awareness among consumers, worldwide, of the existence of agricultural production systems and their products, variously called 'organic', 'biologique', 'ecologico' or 'okologisch'. When asked on their understanding of how these systems differ from conventional production, many would use terms such as 'non-use of chemicals', 'pesticide-free', 'free-range' and 'concern for the environment'. Opinion polls are increasingly suggesting that many of the consuming public would prefer their food to be produced in this manner. Their enthusiasm for such products is somewhat diminished by the premium prices that organic products command. Consumers willing to pay premium prices for organic products understandably expect the organic fruit, flour or chilled ready meal to be the genuine article and presumably trust the label to be an honest declaration. Behind the shopper's instinct lies a considerable mass of knowledge, people, meetings, consultation, draft documents and published legislation at national and international level which has emerged, arguably over the course of this century, but particularly since the 1970s when the first organic standards were published.

This chapter aims to explain the development, current status and future of the standards and regulatory framework that now exist to bring confidence to the shopper's purchase of an organic product, and some of the challenges still ahead. We will concentrate on Europe, as it is furthest ahead in developing regulations and later comment on the position in other countries. Detailed aspects of the certification process are dealt with in Chapter 3.

2.2 The road to legislation

When organic pioneers such as Rudolf Steiner, Albert Howard and Lady Eve Balfour first published their ideas on agriculture in the 1920s, 30s and 40s, it was more an expression of ideology than an attempt to define what biodynamic or organic agriculture was. It is doubtful whether they foresaw the need for detailed legislation which today defines the minimum perch space and type of feed ingredients that allow a hen's eggs to be labelled as organic. Their interest lay in drawing attention to the biological basis of soil fertility and its links with animal and human health. This was at odds with the growing trend, at the time, which considered soil an inanimate medium and placed emphasis on the mineral content as the basis of fertility. Increasing this fertility through the application of soluble fertilisers was becoming the accepted 'scientific' approach to farming.

Arising from such pioneers, disparate farmer groups in parts of Europe, the US and further afield developed their own ideas based primarily on a commitment to a philosophy rather than a market opportunity. Acceptance as an organic producer in the 1940s and 50s was initially simply based on becoming a member of these groups, as such a declaration against the conventional sector was considered a sufficient act of commitment in itself. Informal inspections took place and loose codes of conduct were set out, but the pressure to strictly define organic production systems did not exist because consumer interest was limited to the 'alternative' sector and links between producer and consumer were often close.

The Demeter biodynamic label grew directly out of the teachings of Rudolf Steiner and was probably the first organic label to develop. One of the other early attempts to define organic production came from the Soil Association, the charity that Lady Eve Balfour founded in 1946. The Association published its first standards in 1967 (see Table 2.1) primarily as a means of protecting the consumer and the genuine organic farmer from bogus claims. The preamble to the publication stated that:

'The creation and sustenance of a Living Soil is fundamental to the success of any organic enterprise, and to the quality of its produce. The use of, or abstinence from, any particular practice should be judged by its effect on the well being of the micro-organic life of the soil, on which the health of the consumer ultimately depends.'

Table 2.1 Summary of first organic standards published by Soil Association (source: Soil Association 1967).

Topic	Practice
Soil husbandry	
Recommended	Use of alternate husbandry with complex herbal leys and/or established permanent pasture, and preferably mixed stocking or of varied rotations
	Use of green manure crops with or without sheet composting
	Use of the following organic manures: natural compost and farm manure produced on the farm itself, dried blood, fishmeal (unfortified), feather meal, hoof and horn meal, seaweed meal and liquid extract, pig bristles, shoddy sawdust, bonemeal (unfortified), sheep's trotters, wood shavings
	Use of the following mineral fertilisers: basic slag, limestone chalk, rock phosphate, granite dust
Permitted	If the material is known to be free from chemical and antibiotic residues when applied to the land: battery manure, municipal compost (recognised), farmyard manure (bought in), slaughterhouse waste, mushroom compost, tannery waste, recognised proprietary organic manures e.g. Grancreta and Regenor
Prohibited	All soluble chemical fertilisers
	The continuous growing of cereals
Crop husbandry	
Permitted	Copper fungicides in leaf stage only
	Dispersible sulphur
	Herbal sprays
	Insecticides of vegetable origin such as: Derris, Nicotine, Pyrethrum, Quassia, Ryania, herbal sprays
Prohibited	All other fungicides
	All herbicides are questionable and should not be used on crops to be sold as organically grown
	All pesticides that are persistent, cumulative or toxic to other species
	Most seed dressings are questionable and should not be used on crops to be sold as organically grown, with the possible exception of copper sulphate fungicide

Table 2.1 *(Continued.)*

Topic	Practice
Animal husbandry	
Recommended	That all young stock should have daylight and well ventilated housing, when conditions necessitate their being housed
	That all growing cattle, sheep and poultry should have period of being free range
	That during the finishing period, fattening birds should have daylight, warmth and fresh air
	That feeds should be prepared from organically grown ingredients, supplemented by seaweed, boneflour or other natural minerals, with the inclusion of herbal leys or strips, in the grazing of ruminants
	That calves should have at least 4 days on their dam's milk and 6 weeks whole milk feeding
	That coarse grain should be included in the diet of all laying birds
	Experience has shown that the adoption of the 'recommended comfort and feeding' practices encourages stock to develop a natural ability to surmount disease
	Where, however, disease occurs, herbal treatments should be used in the first instance
Permitted	That calves and young pigs remain housed, provided they have fresh air, daylight, warmth and adequate space
	That fattening cattle and pigs remain housed for the finishing period, provided they have adequate air, daylight, warmth and space
	That hens be kept on deep litter, provided they have reasonable access to grass or a regular supply of green food and natural light
	That feeds should be prepared by the local merchant, provided that the ingredients are known not to include any of the 'prohibited' additives
	The minimal use of drugs and antibiotics in the case of serious illness
Prohibited	The permanent housing of dairy cattle and sheep
	Sweat boxes
	Battery hens and broilers
	Any overcrowding
	The use of antibiotic foods, urea, and other chemical additives
	Veal calf rearing milk substitutes
	The routine use of drugs and antibiotics
	Hormones for chemical sterilisation and caponisation and as growth stimulants
	Debeaking

These principles remain very much the focus of organic management today. Farmers were invited to register their farms with the Soil Association and sign a declaration that they would abide by these guidelines. On-site inspection to verify that farmers met the standards did not commence until the mid-1970s, and with this, the first organic logos were born. At this time the market for organic food was small and there was no interest from trading standards officers or from legislators on what constituted an organic product.

Voluntary standards and inspection systems began to develop independently in parts of Europe, the US and Australia. Their growth and development were organic in themselves, primarily driven by the producers and concerned consumers. Many of the early certification programmes developed as producer/consumer groups and some (Soil Association, California Certified Organic Farmers) retain this balance today.

With the increasing global interest in organic agriculture came the birth of an international movement brought together under the International Federation of Organic Agriculture

Movements (IFOAM) in 1972. Its mission was to enable exchange of information and ideas and to foster co-operation across cultural, language and geographic barriers. IFOAM published its understanding of Organic Standards in 1980 and has continued to revise them on a biennial basis ever since (see section 2.6.1).

Until well into the 1980s, governments took little notice of the developing organic movement, generally considering it to be a cranky sideshow to the real business of 'agriculture based on science'. As organic products began to appear in more mainstream retailers in Europe and the US in the 1980s and trade started to increase across borders, the authorities became more interested in the regulation of the market and concerned at the potential for fraudulent claims and confusion in the consumer's mind concerning what constituted 'organic'. Legislation became necessary. Figure 2.1 charts the progress from ideology to legislation and the factors that influenced the developments.

2.3   Standards for production systems, not products

The core challenge to the pioneer farmer and consumer groups, which remains to this day, is that it was, and still is, not possible to test a product's organic integrity by measurement or analysis. Absence or presence of chemical residues cannot confirm that the product was produced under organic or conventional management respectively. Though some research work has indicated dry matter differences between organic and conventional potatoes for instance, and German scientists have claimed crystallography techniques can distinguish one from another (Balzer-Graf & Balzer 1988), these techniques remain impractical and too esoteric for many. The approach to defining and policing organic products that emerged, and which now forms the basis of all organic standards and the regulatory system that exists today, was to set standards for the method of production rather than the final product. This was, and remains, a novel and challenging approach to standardisation and verification in that it necessitated registration and assessment of farms long before any product reached the market. It also required detailed definition of the production practices allowed or disallowed. The 'no chemicals' understanding of the consumer was clearly not enough. The principles were defined by the various producer organisations through consultation with their members and, somewhat characteristically, resulted in splits in the movement, which led to a number of different standards being developed within the same country, let alone across the world.

Given the complexity of farming systems and the wide variation in agro-ecological and social conditions that influence them, this seems hardly surprising. It is perhaps more surprising that at the end of the 1990s, there is, globally, a broad understanding and agreement of what constitutes organic food production and processing. IFOAM, a non-governmental organisation, can largely be credited with this achievement, seen to be representing, as it does, the organic movement worldwide. IFOAM's Basic Standards and the IFOAM Accreditation Programme (see section 2.6.1.1) are generally respected as the international guideline from which national standards and inspection systems may be built and have been used extensively as a reference by standards setters and legislators. Chapter 3 provides more detail on inspection and certification.

Notwithstanding this, there remains much argument over details of standards, operating procedures for certification programmes and the procedures for determining equivalence between products produced in different parts of the world.

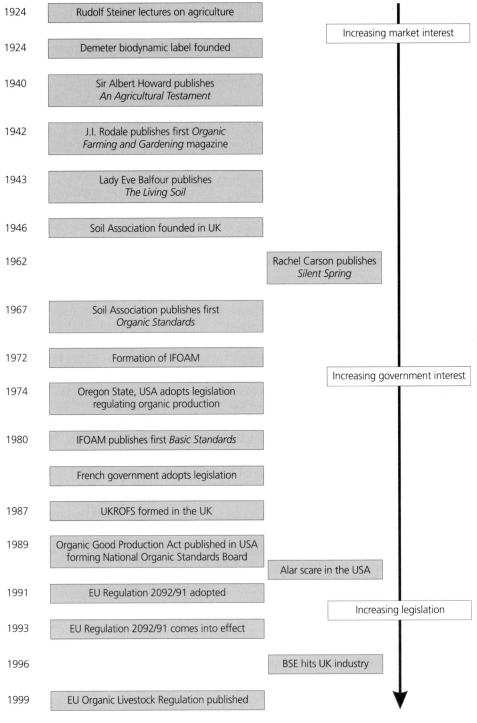

Fig. 2.1   The organic movement: progress from ideology to legislation.

2.4   Legislation

Although the adoption of Regulation 2092/91 (European Commission 1991) across Europe in 1991 covering the labelling of organic foods was not the first piece of legislation to be developed in the world (France, Spain and Denmark already had national legislation as did some US states), it has probably had the most far-reaching consequences to date on the organic movement. This impact has arisen through the combined circumstances of being the first regional, statutory definition to appear and that Europe represents one of the largest markets for organic produce. Businesses both inside and outside of Europe had to comply if they wanted to sell within or into the European market. As we shall see, the impact of Regulation 2092/91 has been both positive and negative within Europe and outside.

2.4.1   *European Regulation on organic agriculture*

The initial format and content of the Regulation 2092/91 published in 1991 was developed over two years and set out timetables and structures for review and allowed for amendments. The Regulation covered only unprocessed and processed crop products initially and left a number of areas open for addition and review, most notably the detailed rules on imports and those covering livestock. Both topics have been subjected to much discussion throughout the 1990s (see below). The guidelines for wild crop products and mushroom production were added later.

Crop products not intended for consumption, such as herbal remedies for external use and manufactured cotton textiles, remain outside the scope of the Regulation. Wine also remains outside the scope. This leaves a grey area that is not covered by legislation but where some private certification bodies inspect and certify and allow 'organic' labels to be applied. It is likely that all of these categories will be brought under the Regulation in future. As indicated below, where the Regulation does not cover a product, it refers to national or internationally accepted rules.

Aims

The rationale for the Regulation is set out in the recitals of 2092/91at the beginning. *At the risk of some inaccuracy, the authors have avoided the legal language of the Regulation and tried to summarise the intention of the text for clarity.* It recognised:

- The developing market for organic foods
- The price premium for organic foods
- The environmental benefits offered by organic production
- The previous existence in some Member States of standards and inspection systems.

It proposed that the Regulation would:

- Ensure conditions of fair competition between organic producers
- Improve transparency and understanding of organic production systems, and
- Improve credibility of organic products in the eyes of consumers.

It then set a framework for the regulation, which included:

- That organic production can be well defined and minimum standards should be set down
- Provision needs to be made for the regulatory framework to be flexible and amendable
- Any unclear aspects of the Regulation should refer to national or international guidelines
- The provision to add requirements on prevention of contamination
- The need for verification of the complete production and handling chain
- A system of inspection and a community-wide logo.

Structure

Figure 2.2 provides a summary of the contents of the Regulation including its amendments. For ease of understanding, the articles and annexes have been broadly separated into categories relating to inspection arrangements, labelling requirements, production rules and administration of the Regulation itself.

Implementation

The main impact of the Regulation was to change the need for registration of operators involved in growing, preparing and selling organic products from a voluntary system to a legal requirement. Any product labelling claiming any of the designated indications (organic, *biologique*) meant that the product must have been produced, packed and/or processed by an operator or chain of operators who had been registered with an approved inspection body. Production rules governing management on the farm, and in the food factory particularly, were relatively basic and relied heavily on lists of allowed inputs in the former and ingredients and processing aids and permitted ingredients of non-organic origin in the latter.

The Regulation (Annex III) set out a minimum frequency of on-site (farm, factory premises or warehouse) inspections of one per year and specified the information that was to be retained by the operator and verified by the inspection body to determine compliance. In some instances the level of record-keeping was increased from the previous voluntary systems in place and some operators have complained at the additional burden this imposes. The burden was felt most by small growers, and this, along with the increasing cost of certification, has led to many small enterprises leaving the certification system.

Article 9 of the Regulation set rules for implementation of the inspection system and requires that each member state set up a competent authority to implement or to approve and oversee private inspection bodies performing the work. Finland, Denmark and Spain have opted to implement the service through government departments. In all other member states government competent authorities have approved private inspection bodies to do the work. Government inspection bodies were unable to require stricter inspection measures than that set by the Regulation, whereas they can approve the operation of inspection bodies with higher requirements. This requirement has since been changed by the Livestock Regulation (for livestock rules only) allowing stricter measures to be maintained on a national level (but this cannot be used as a barrier to trade in the context of the single market). In the UK, the competent authority (UK Register of Organic Food Standards

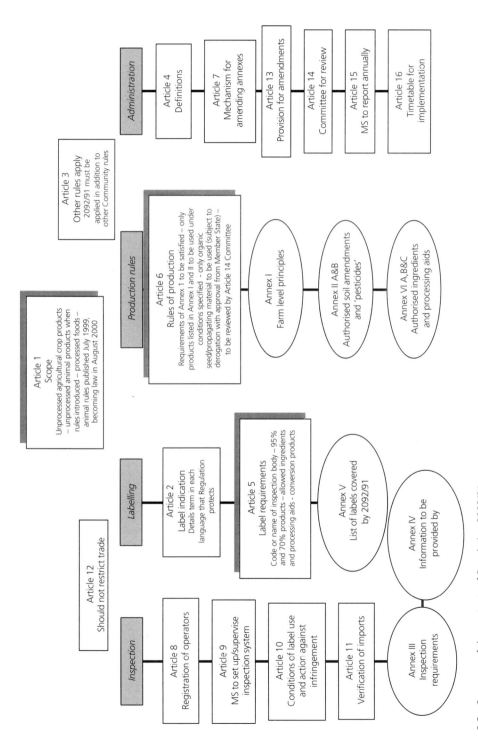

Fig. 2.2 Summary of the contents of Regulation 2092/91.

– UKROFS) oversees six inspection bodies, several of which implement their own standards and others follow the Regulation exactly. UKROFS also operates as an inspection body itself.

At the level of the inspection body, the Regulation specified the need for procedures, competence, independence of decision making, appropriate resources and skills and data protection. It is the responsibility of the competent authority to verify an inspection body's performance. This is generally done by an inspection of the inspection body and also repeat monitoring inspections of a sample of their registered operators. Since 1 January 1998 the initial approval and ongoing verification of competence of an inspection body is guided by compliance with the European Norm, EN45011, the EU version of ISO65, which sets guidelines for the operation of certification bodies. Accreditation to this norm was not a requirement of the Regulation although some competent authorities (France) have required it of their respective inspection bodies. EN45011 has however caused much greater concern outside of Europe in relation to assessing equivalence of imports (see section 2.5.2.3 below).

Amendments

The initial Regulation was a far from perfect document leaving many areas open to interpretation. Whether adopting the Regulation, even in that form, was a better approach than the agonising delays that have taken place in the US (see below) is open to conjecture. Amendments and many timetable adjustments have taken place, generally several times per year, since 1991. The sequence and subject of the Council and Commission Regulations are summarised in Table 2.2 and give a flavour of the main omissions and, perhaps, the over-optimistic timetabling that was initially set.

Many of the amendments have been prompted as a result of practical experience and continuing discussion between member states. The increased definition of the composition and conditions of use of soil amendments and pest and disease control products forming Annex II is a good example of the increasing level of detail that the Regulation has embraced. It is well known that the initial intention of the Commission when commencing the development of the Regulation was a fairly basic statement of the need to avoid the use of synthetic chemicals. It soon became clear that much more detailed rules would be required.

Livestock Regulation

The long-awaited rules of production for livestock (Council Regulation (EU) No. 1804/1999) were approved by the Council of Ministers on 19 July 1999 and published in the *Official Journal* on 24 August. Their impact and interpretation by member states and inspection bodies are still (at the time of writing) being analysed. The Regulation will enter into force from 24 August 2000. Significantly, the Regulation also specifically prohibits the use of genetically modified organisms and their derivatives, and this requirement entered into effect immediately.

Regulation 1804/99 amends and supplements 2092/91, adding in the necessary references and rules for livestock. Some of its main components are summarised in Table 2.3 (for a comprehensive view, readers are advised to consult the Regulation itself). This Regulation also updated and had impact on issues relating to crop production. One major change is the limiting of applications of manure to a maximum of 170 kg N/ha/year over the whole farm (aimed at ensuring that livestock production was 'land based' and therefore properly integrated with the farming system).

A further change impacting upon intensive horticultural holdings is that brought-in manure or proprietary manure products must now be derived from organic stock or stock managed extensively and shown not to be fed genetically modified ingredients. In many regions, this appears to threaten the existence of organic protected cropping, other intensive field vegetables and mushroom production.

Amending the Regulation

A small organic unit of Directorate General VI (DGVI, now called Directorate General Agriculture) of the European Commission (one full-time senior staff member only) have had the task of co-ordinating the implementation and revision of the Regulation. The so-called Article 14 Committee, made up of delegates from each member state competent authority, meets regularly to negotiate and discuss papers prepared by the Commission. It is intended that the Article 14 Committee members consult widely within their own countries, but it is clear that the level of consultation by Committee members varies between member states.

The Commission will also, on occasions, bring in consultants to review the operation of certain aspects of the Regulation, as was done in 1998/99 on the mechanism for inspection bodies and competent authorities to handle the verification of imports.

Issues raised by the EU Regulation

Overall most operators and inspection bodies would accept that the Regulation has brought some improved clarity and uniformity to the organic market. However, differences of interpretation have inevitably arisen at the member state level, leading to calls from some for additional guidance. At the same time others claim that there is over-regulation.

Too much detail, however, may lead to conflict between countries and regions with as diverse agro-ecology and social structures as Greece, Germany and Finland. This has led to the introduction of the phrase – 'need recognized by the inspection body' – which essentially devolves the decision to the member state and inspection systems that operate there. This is generally seen as the only sensible way to deal with regional variation, but can also be at the root of allegations of unfair competition when different interpretations may be used by inspection bodies and competent authorities on the appropriate use of an input. This especially arises when products approved by an inspection body in one country are sold in another (see section 2.5 below).

There also remains concern in the organic movement over who is setting standards and the level of involvement of the industry itself. The Article 14 Committee, for example, is made up solely of public sector members. The EU Group of IFOAM, as representatives of the organic industry in the EU, are given special consultation access to the DGVI Organic Unit twice a year but no official minutes are published of these meetings.

The established positive lists for soil amendments, pest products and processing aids and so on, and the length of time it takes to amend these lists, is also a concern. The Commission undertakes long periods of consultation with member states to ensure the right balance between stringent standards, the expectations of consumers and what is achievable under current circumstances. However, availability of organically grown

Table 2.2 Amendments to Regulation 2092/91 since publication.

Date of amendment	Reference	Subject	Main points
14 January 1992	Commission Regulation (EEC) No. 94/92	Import	Describes mechanism for approving a country's entry onto the Third Country list, which accepts production rules and inspection systems as equivalent to those of the EU. The preliminary list was repeatedly amended and implementation delayed. Finally implemented March 1997
15 June 1992	Commission Regulation (EEC) No. 1535/92. Integrated as statement on animal products in Annex I and additions in Annex III	Animal ingredients and operator measures	Until detailed rules on animal production are developed, animal products used as minor ingredients should comply with national/international rules. Minor additions to inspection requirements including organic integrity check of product delivered to a processor
14 July 1992	Council Regulation No. 2083/92 EEC. Integrated as Article 11.6 and amended Article 16.3	Imports and date of implementation	Allowed a third option based on a case-by-case assessment of equivalence by the competent authority in the Member State. Delayed implementation of requirements for labelling, inspection and import equivalence until 1 January 1993
30 November 1992	Commission Regulation (EEC) No. 3457/92	Import certificates	Provides rules for who issues import certificates for products arriving from Third Countries, their format and use
22 December 1992	Commission Regulation (EEC) No. 3713/92	Delay to implementing the Approved Third Country list	See Regulation 94/92 above. The approval process for determining a country's production rules and inspection system to be equivalent took longer than anticipated. Superseded
29 January 1993	Commission Regulation (EEC) No. 207/93. Integrated as Annex VI	Use of ingredients of non-agricultural origin, of conventionally produced ingredients and processing aids	Established Annex VIA, B and C which are limited lists of ingredients and processing aids that can be used in organic products. Provides guidelines and mechanisms for amending it and how operators may obtain derogations for limited periods
24 June 1993	Commission Regulation (EEC) No. 593/93	Delay to implementing the Approved Third Country list	Further delays in approving equivalence
23 September 1993	Commission Regulation (EEC) No. 608/93. Integrated into Annex II and III	Wild products, use of calcium chloride and inspection rules for importers	Introduction of rules and inspection requirements for wild harvested products, the addition of calcium chloride to Annex IIA and rules for inspection of importers
2 March 1994	Commission Regulation (EEC) No. 468/94. Integrated into Annex VI	Addition of ingredients/processing aids to Annex VI	Added *laungengebaeck*, rice meal and lactose to permitted lists
28 March 1994	Commission Regulation (EEC) No. 688/94	Defer implementation of Approved Third Country list	Further delays in finalising list

Date	Regulation	Subject	Description
20 June 1994	Commission Regulation (EEC) No. 1468/94	Conversion labelling	Extended validity of use of conversion labelling. Superseded by 1935/95
30 September 1994	Commission Regulation (EEC) No. 2381/94 New Annex IIA introduced	Changes to soil amendments allowed and their conditions for use	Introduced a much more detailed Annex IIA as a result of recognising the need to specify origin, composition and use of the various soil amendments
24 October 1994	Commission Regulation (EEC) No. 2580/94	Amendment to proposed Third Country list	Removal of an Argentine inspection body from proposed list based on on-the-spot-evaluation. Superseded
9 March 1995	Commission Regulation (EEC) No. 529/95	Defer implementation of Approved Third Country list	Further delay in implementing the Third Country list and minor change to the import certificate
29 May 1995	Commission Regulation (EEC) No. 1201/95	Amendments to Annex VI	Based on changes in availability of organic ingredients, additions and deletions were made to Annex VIC
29 May 1995	Commission Regulation (EEC) No. 1202/95. Integrated in Annex I and III	Special circumstances for reduced conversion and parallel production	Reduced conversions possible where mandatory pesticide use required and parallel production allowed for certain situations with precautions
22 June 1995	Council Regulation (EC) No 1935/95	Overhaul of Basic Regulation	Extension of conversion labelling period, delay in developing animal regulation, labelling, restrictions on use of conventional propagating material and seeds, changes to soil amendments list and extension of expiry of importer derogation provision. Introduced requirement for inspection bodies to comply with EN45011
7 March 1996	Commission Regulation (EC) No. 418/96	Amendments to Annex VIC	Allowed the use of certain conventional spices due to inadequate supplies of organic
26 March 1996	Commission Regulation (EC) No 522/96. Amends 94/92	Import equivalence	Modified Third Country list of 94/92, delayed implementation until 1 January 1997. Superseded by 314/97 below
20 February 1997	Commission Regulation (EC) No 314/97. Amends 94/92	Import equivalence	Indicates finalisation of evaluations and presents Modified Third Country list including addition of a Dutch inspection body as an approved body in Hungary
26 February 1997	Commission Regulation (EC) No. 345/97. Amends 207/93	Use of conventional ingredients	Allowed for repeat derogations on the use of conventional ingredients and added more requirements for information supplied
29 July 1997	Commission Regulation (EC) No. 1488/97	Amendments to Annex IIA and B and VIA and B	As a result of representations from Member States a number of additions were made to permitted soil amendment and pest product lists and agricultural ingredient, processing aid and conventional product lists
29 June 1998	Commission Regulation (EC) No 1367/98	Amendments to 94/92	Additions/changes to lists of approved inspection bodies
4 September 1998	Commission Regulation (EC) No 1900/98	Addition of rules on mushrooms	Substrates for mushroom production must be from organic farms. Derogation to allow up to 25% of substrate from extensive agriculture with reference on label
19 July 1999	Commission Regulation (EC) No 1804/99	Livestock Rules	Livestock production rules added to 2092/91 to come into force on 24 August 2000. Prohibitions on the use of GMOs effective immediately

Table 2.3 Summary of main points of Livestock Regulation 1804/99 (source: European Commission 1999).

Reference	Subject	Issues
Article 1	Scope	Adds livestock (includes bees but not aquatic livestock), unprocessed livestock products, processed livestock products and a timetable for labelling of animal feeds (24/8/2001)
Article 2	Indications	Restates indications in each language and adds diminutives such as 'bio' and 'eco' but allows a transitional period (1/7/2006) for trademark holders to adapt
Article 4	Definitions	Products of hunting and fishing of wild animals/fish cannot be considered organic
Article 5	Labelling	Declaration on 'produced without the use of GMOs …'
Article 6	Production rules	Allows the use of veterinary products derived from GMOs. Delays the requirement for organic seed use until 31/12/2003
Article 12	Free trade	Allows for adoption of more stringent rules for livestock products by Member States for those produced in their own territory
Article 13	Future changes	Allows provisions for amending Annexes, the development of Community logo, the use of GMO derived veterinary products, possible move to a threshold for GMO contamination
Annex I	Principles	Must be land based – prohibits landless production – derogations possible Prohibits raising of conventional stock on same land or unit – derogation possible Allows use of common land provided it has been untreated for 3 years
	Conversion	Derogation possible reducing pasture land conversion to 12 or 6 months Animals kept for meat can be converted – sets minimum periods under organic management Simultaneous conversion of land and livestock possible
	Origin of animals	Livestock existing on the holding can be converted Sets maximum ages of young stock when brought in during establishment Availability of organic stock will be reviewed before 31/12/2003
	Feed	Up to 30% feed can come from in-conversion sources Derogation up to 2005 that allows up to 10% for ruminants and 20% for non-ruminants can be conventional on annual basis Only ingredients listed in Annex II can be used
	Disease prevention	If animal becomes sick, must be treated immediately Prohibits use of preventive allopathic veterinary products Withdrawal period twice legal period Limit on number of allopathic treatments per year
	Husbandry	Artificial insemination allowed Castration allowed with conditions Tethering forbidden but with derogation for existing buildings until 2010
	Manure	Limit of equivalent of 170 kg N/ha can be applied per year, though co-operation with other holdings is allowed. Annex VII sets maximum stocking rates Sufficient storage capacity required
	Housing	Minimum indoor housing and exercise areas set down – Annex VIII Permitted cleaning products are specified in Annex II Pasture access required though may be waived under certain conditions Maximum house populations for poultry specified General housing derogation up to 2010 where buildings existed before 8/99
	Bees	Conversion over one year Bees must originate from organic colonies – derogations possible Foraging area of 3 km radius on organic land or 'spontaneous vegetation' Distance from urban and industrial areas left to Member State to establish Artificial feeding allowed with conditions

products and new pest and soil products are emerging constantly, but the lists cannot be amended rapidly.

For different reasons, some traditional products used in organic systems, such as neem insecticide, have only restricted use under the EU Regulation, as they are not registered as a pest product within the EU.

2.4.2 Other countries

The status of legislation governing the production and labelling of organic foods in different countries around the world varies widely. The majority of countries have no legislation and have no programme developing such a framework, whereas others are actively developing legislation. A last group are those that responded quickly to the Regulation developing in the EU; Argentina, Australia, Hungary, Israel and Switzerland all developed legislation closely in line with the EU Regulation, allowing them to be evaluated and approved as operating equivalent systems. Table 2.4 summarises the current status of legislation and organic certification provisions.

The absence of legislation does not imply that organic production is less stringent or that the market development is necessarily impaired. Neither the US nor Japan (see below) have implemented organic legislation, but are two of the world's biggest markets for organic food. Both, however, have been working on legislation over the last ten years and see it as crucial for further sustainable market development and protection of the consumer. In countries where no specific organic food legislation is in place, the market may be regulated by other labelling laws, such as in Canada. In many less developed countries, foreign certifiers provide inspection and certification services to operators in a more or less unregulated environment.

USA Organic Rule

Alongside the developments in Europe a parallel sequence of events has unfolded in the US beginning with the development of production standards by a farming organisation called California Certified Organic Farmers in 1973. Other farmer organisations and some States drafted their own standards and legislation, not necessarily with reference to each other, again resulting in variability across the country. There are currently over 30 private and 11 state organic certification agencies operating in the US. As a result there is a mixture of unregistered farms and operators selling foods labelled organic, other operators who may be registered with their State system but with no formal inspection process, through to operators who subject themselves to comprehensive inspection systems run by private certifiers accredited by IFOAM (see section 2.6.1).

In the 1980s the Organic Trade Association attempted to establish a national voluntary organic certification programme but was unable to come to a consensus on standards. Congress was then petitioned by the industry to establish a mandatory organic programme. In 1990 the Organic Food Production Act was published which set out to:

- Establish national standards governing the marketing of certain agricultural products as organically produced products
- Assure customers that organically produced products meet a consistent standard
- Facilitate commerce in fresh and processed food that is organically produced.

Table 2.4   Status of legislation and inspection services worldwide.

Region/Country	Status of legislation	Inspection services
AFRICA	None	Foreign certifiers/some projects to develop local capacity
ASIA		
China	None	Domestic and foreign certifiers
Malaysia	None	Foreign certifiers
Papua New Guinea	None	Foreign certifiers
Thailand	None	Foreign certifiers/developing domestic
EUROPEAN UNION	Implemented	Domestic certifiers
REST OF EUROPE		
Czechoslovakia	Implemented	Domestic and foreign certifiers
Hungary	Implemented	
Poland	None	Domestic and foreign certifiers
Switzerland	Implemented	
Turkey	None	Foreign certifiers
MIDDLE EAST		
Egypt	None	Domestic and foreign certifiers
Iran	None	Foreign certifiers
Israel	Implemented	
FORMER SOVIET UNION		
Lithuania	None	Domestic certifier
Russia	None	Domestic and foreign certifiers
Ukraine	None	Foreign certifiers
NORTH AMERICA		
Canada	Draft	Domestic and foreign certifiers
USA	Adopted but not implemented	Private and state certifiers
SOUTH AMERICA		
Argentina	Implemented	
Belize	None	Foreign certifiers
Bolivia	None	Domestic private certifier
Brazil	None	Domestic private certifier
Chile	In process	Domestic and foreign certifiers
Costa Rica	None	Foreign certifiers/developing domestic
Mexico	None	Domestic inspectors/foreign certifiers
Paraguay	None	Foreign certifiers
PACIFIC RIM		
Australia	Implemented	Law only mandatory for exports
Japan	In process	Domestic and foreign certifiers
New Zealand	Implemented	Domestic certifiers

Under the OFPA a National Organic Standards Board (NOSB), made up of organic farmers, handlers and retailers, various experts in environmental protection and consumer interest groups was appointed by the Secretary of Agriculture in 1992. The NOSB formed specific working committees on crops, livestock, processing, packaging and labelling, materials, accreditation and international, and developed position papers inviting public comment. After a period of consultation the NOSB presented its recommendations to the Secretary of State in August 1994. Based on this submission the Agricultural Marketing Service of the USDA published its Proposed Organic Rule in December 1997. Despite the apparent level of consultation this proposal met strong resistance (over 280 000 comments were received by USDA) from the organic movement, not only within the US, but also worldwide. Some of the most contentious issues were as follows:

- Irradiation not prohibited
- GM ingredients not prohibited
- Use of human sewage not prohibited
- Use of antibiotics not limited
- Pasture access for livestock not required
- Prohibition of private organic labels.

As a result of the reaction to the Proposed Rule, the USDA withdrew to reconsider. In October 1998 it published three National Organic Programme Issue Papers, two on livestock and one on the authority of private certifiers to decertify operators. Once more, significant response was received and no further papers have been produced. As at December 1999, the second version of the Rule has not appeared.

In response to these developments, the Organic Trade Association in the US has once more returned to developing its own independent national standards. A third draft of the American Organic Standards is being prepared at the time of writing. The aims of the American Organic Standards, as stated in draft 2 (September 1999), are to:

'Encompass principles of organic production for crops, livestock, processing, handling, and labelling, and establish certification and accreditation criteria.

'Harmonize standards for the production, processing, handling, labelling, and certification of organic plant and animal products.

'Establish a set of standards and certification and accreditation guidelines supported by the United States organic industry and community, including existing public and private certification agencies.

'Provide a set of standards in operating manual format that can be used by certification agencies to protect consumers and producers against deception, fraud, and unsubstantiated product claims in the market place, including misrepresentation of other agricultural products as being organic.

'Provide a set of certification and accreditation guidelines useful for the establishment and administration of an accreditation program for organic certification agencies.

'Enable producers, processors, handlers, inspectors, and certification agencies to assess an operation's compliance to a uniform organic standard.

'Provide a baseline standard for certification of growers, handlers, and processors as well as guidelines for the accreditation of US-based organic certification agencies without precluding certification agencies or governments from creating more restrictive standards.

'Provide a baseline standard that will result in reciprocity between certification agencies accredited according to these standards and provide the basis for the negotiation of additional standards recognition agreements between accredited certification agencies.

'Facilitate equivalence with international guidelines for organic certification, accreditation and labelling in order to encourage local, regional, and international trade in organic products.'

The thinking behind the industry standards is to build consensus within the industry in preparation for the next comment period on the Regulation, which is expected to be released by USDA towards the end of 1999. It is true to say that the US and the wider world organic movement await the next USDA draft with interest.

2.4.3 *Japan*

There is no legislation governing organic food and farming in Japan, but at the time of writing, the Japanese government is in the process of drawing up a law. It is unclear precisely how this is intended to operate.

2.5   International trade and equivalence

Considering the problems the organic movement and the legislators have had in various countries to come to some agreement on standards on a national level, it might seem a tall order to achieve agreement internationally. Efforts on harmonisation are considered in section 2.6 below, as these are a considerable topic of debate and development. Past and existing approaches to imports from countries not subject to the same or any statutory regulations have taken the approach of determining equivalence of standards and inspection procedures.

2.5.1 *Private certifiers*

The growth of organic foods into an international business has meant that organic processors and importers seek far and wide for products and ingredients, which may be certified by any one of over 100 certification programmes operating in the world. Despite attempts at harmonisation, discussed below, standards and procedures vary. This has necessitated that all certifiers establish some mechanism for evaluating the equivalence of the standards and procedures of other certification bodies. This is generally done in one of two ways:

(1) Case-by-case evaluation of inspection reports, standards and operating manuals for each product imported. Questions arise as to how far one traces back the various ingredients of a multi-ingredient product.
(2) Overall evaluation of a certification programme's operation leading to a blanket acceptance or specific category exclusions where necessary.

In countries that rely on high levels of imports such as the UK, or domestic transactions in the US where there are a high number of certifiers, both approaches require excessive levels of skilled labour for any individual certification office. This problem is accentuated by the existence and strength of private organic logos, which are particularly prevalent in Europe.

Private labels

In many countries, clear leading 'organic brands' have developed, such as KRAV in Sweden, the Soil Association in the UK, and the French government-sponsored generic

AB label, each with their own requirements. There is little doubt that each brand will fight to maintain market share, in part to maintain the viability of the inspection body, but also to further their own view of what constitutes organic.

The outcome is that, even for domestic or sales within Europe, operators are, for market reasons, pressured to subject themselves to a recertification process by the inspection body which dominates the local market. There is no denying of access to markets, so there is no non-compliance with the Regulation. The Regulation set minimum production rules and operating procedures, which superseded any national legislation in any member state. It did not outlaw the existence of private certification providers who can set their production standards at or above (but not below) the Regulation level. So although the intention was to improve understanding and clarity, there are still differences between production standards set by different inspection bodies, albeit minor. This has led to the inspection body of the buyer re-evaluating the inspection and documentation of the inspection body relating to the seller and making its own decision on compliance with its own standards. This is acceptable within the legal scope of the Regulation but probably not within the spirit of what the Regulation set out to achieve. It also amounts to a lot of hard work for the operators and the inspection bodies. So why does it continue?

Many of the older, longer-established inspection bodies existed before the Regulation came into effect and originated out of campaigning organisations. Some have hard-earned reputations of integrity and high standards that they do not wish to lose and also had, and retain, stricter requirements on operators than those brought in by the Regulation. Their position, which in some circumstances is reinforced by their Accreditation by IFOAM (see section 2.6.1.1), is to maintain upward pressure on standards ahead of the inevitably more general requirements of the Regulation. This creates choice for operators and for consumers but also some potential confusion for consumers and fragmentation of the organic industry. Circumstances arise in which an operator within a country could select to be certified by one body which has lower standards than another. The product can still be labelled 'organic'. The differences tend to arise in those areas in which there are accepted/allowed uses of conventional approaches under certain circumstances such as proportion of conventional feed permitted, the circumstances of using conventional medicines like anthelmintics or the use of copper fungicides.

As the market grows, change in this area is inevitable. Whether it will be a fundamental change in the way programmes view equivalence or just streamlined procedures for assessing equivalence, remains to be seen.

2.5.2 *Impact of the EU Regulation on international trade*

Within the context of the EU Regulation, the approach to assessing equivalence of third country (i.e. countries outside the EU) exports and their inspection systems are much the same as that developed by private certifiers.

Trade within the EU

The Regulation clearly states in Article 12 that member states cannot restrict movement of organic products within the Community. Essentially any product from a registered producer operating within a member state, which complies with the Regulation, can be sold in any other member state. Finished products produced in any member state can be sold to any trader in Europe, registered or not. Transactions involving bulk products or products intended for processing or repackaging must take place between registered operators. For example, an olive oil produced and processed by a registered operator in Andalucia can apply an organic label to the product and ship it to Germany for sale in a supermarket. No verification by the German authorities is required unless the oil is supplied in bulk for reprocessing in Germany. In the eyes of the Commission, this is not considered import – the oil has been produced in the EU.

Therefore in any store selling organic goods, the consumer will be presented with products labelled organic but with different labels. To avoid confusion the Regulation has allowed for the development of an EU logo, with the aim of providing clear recognition of authenticity. The logo would be optional and is yet to be launched, and its reception and impact is uncertain, especially where private labels are strong.

Imports from outside Europe

The regulation of organic integrity of imports from outside the EU has proven to be a sensitive subject and the EU Commission are currently consulting member states on improving the mechanisms available, which, at the same time, reduce fraud and ensure equivalence of standards and inspection, but remain practical.

Operators in countries outside of Europe wishing to supply an importer in any EU member state must be able to demonstrate that:

'the product was produced according to organic production rules equivalent to EU standards;

'the production process was subject to inspection measures equivalent to the EU inspection requirements; and

'that the inspection measures will be permanently and effectively applied'
(Council Regulation (EEC) No. 2092/91, Art. 11 para.6, as amended)

In addition, a consignment must be accompanied by an 'EU Certificate for the Import of Products from Organic Production', sometimes called a transaction certificate. The original must accompany the shipment to the first consignee and be kept by the importer for two years. The inspection/certification body who verified the operation of the exporter issues the certificate.

There are three routes to complying with the above requirements:

(1) EU Commission evaluation of a third country's system for accrediting private inspection or certification bodies.
(2) Assessment on a product by product basis known as the 'importer derogation'.
(3) EU member state evaluation of a third country's inspection/certification body.

EU Commission evaluation and acceptance as an Approved Third Country. This was the original method that the Regulation 2092/91 envisaged for the approval of imports from outside the EU member states. However, approval requires the existence of legislation and considerable industry collaboration and resources to enable submission. This has deterred or prevented many countries from applying, so that, as of July 1998, only five countries had been approved and appear on the list that forms the Annex to Regulation 94/92 as amended.

The application to the Commission must include:

- The legislation covering organic standards and operation of the inspection system.
- If relevant, the mechanism to approve and supervise private inspection/certification bodies.
- The types of products to be exported.
- Evidence that inspection/certification bodies operate in compliance with EN45011/ISO65 (see below).

The Commission conducts on-site evaluation of the regulatory system including sample operator inspections before approval. Currently, the countries approved by this mechanism are Argentina, Australia, Hungary, Israel, Switzerland and the Czech Republic.

Importer derogation. The competent authorities can authorise the import of products from countries outside the EU on a case-by-case basis, subject to receiving an application from an importer registered with one of the approved inspection/certification bodies. This is the responsibility of the importer, not the exporter. Though the importer derogation provision is set to expire on 31 December 2002, it is expected that it will be made permanent. To satisfy the requirements, the application should normally be accompanied by:

(1) The standards of the inspection/certification body that inspected the producer in the country of export.
(2) The operating manual of the inspection/certification body.
(3) A statement from the inspection/certification body indicating that the inspection measures will be permanently and effectively applied.
(4) Written evidence (from government or an independent accreditation body) that the inspection/certification body operates in compliance with EN45011/ISO65.

Where the competent authority have received previous applications involving the same country/certifier combination, they may not require the above documentation. Clearance of a product from a new country and/or new certification body may take several months.

This route presently accounts for the majority of imports into member states, with Germany handling the greatest number, followed by the Netherlands.

EU member state evaluation. An amendment to Council Regulation 2092/91 (Article 11.7) allowed for an EU member state to evaluate a third country's inspection/certification body and request the Commission to approve it. If the Commission approves, it can then be added to the list published as part of Regulation 94/92 and its amendments (essentially an approved list of countries and certification bodies). This provision was to establish a mechanism under which certification organisations based in EU countries could be approved for certifying imports in Third Countries for import into the EU. It does however allow for third country-based certification organisations to obtain approval.

This route can be considered a medium-term option, negating the need for the third country to implement its own legislation, but allowing the blanket acceptance of products certified by a specified inspection/certification body within an exporting country. Having said that, there has to date been no applications via this route, presumably because it is not the priority of individual competent authorities to promote any particular third country's approval.

Figure 2.3 summarises the various routes that may be taken by exporters in different situations to obtaining the appropriate clearance to allow supply into the EU.

EN45011

When the EU Regulation was first published, it set very few guidelines on the qualifications of certification bodies (both within and outside of Europe) and the procedures for their approval. Regulation 1935/95 added the requirement for all inspection bodies to comply with the guidelines set down in European Norm 45011, which is the EN version of ISO65.

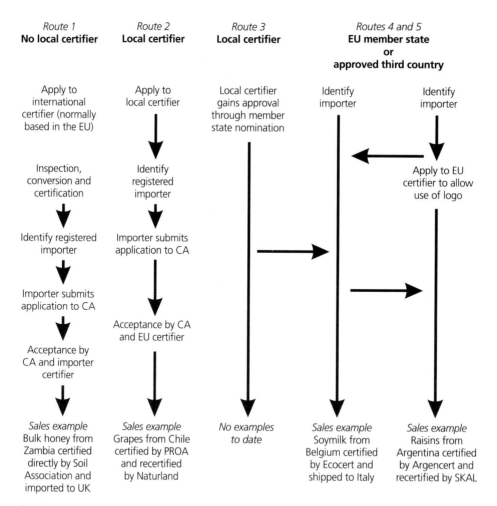

Fig. 2.3 Possible routes to obtaining clearance for supply of organic products into the EU (CA = Competent Authority).

The Norm describes criteria for certification bodies operating product certification. This requirement has caused considerable argument on several fronts:

Compliance with the Norm is required, not necessarily *accreditation*. Some inspection bodies within Europe have opted or been required by their national authority to gain accreditation at considerable expense. In other countries, competent authorities have conducted compliance inspections themselves.

For inspection bodies in third countries, demonstrating compliance can be achieved either by accreditation, government verification or by submitting documents to the member state competent authority to evaluate compliance. In many situations, accreditation is prohibitively expensive, the third country government is unwilling or unable to provide verification, and member state competent authorities are unwilling to undertake such a task. This has left many third country inspection bodies from those based in the US to those in Chile at risk of being unable to provide access to Europe to their operators.

EN45011 is a guideline for product certification, not process certification. Its appropriateness for organic agriculture is in some doubt.

After some delays in implementation from September 1998 to June 1999, the requirement is now in place. The US government was forced at the last minute to introduce a system to verify EN45011 compliance of organic inspection bodies so as to provide access to the European market. The situation is less clear in other countries, but it is generally believed that member state competent authorities are interpreting the legislation in a flexible manner.

2.5.3 *Imports into other countries*

Alongside Europe, the US and Japan are the other major importers of organic products. Most other countries, for example, Australia, New Zealand, Israel, Argentina and Egypt, are predominantly exporters. Legislation governing requirements for imports for this group are either non-existent or present in rudimentary form and untested. As we have seen for Japan, no legislation has yet been implemented, so imports take place largely based on private sector verification, often by the importing companies themselves.

In the US, no federal legislation is implemented, so the situation varies from state to state. However, both the USDA rule and the American Organic Standards (referred to in section 2.4.2.1 above) proposed arrangements for evaluating imports. In California, at the present time, foreign inspection bodies can apply to the state authorities for approval allowing the sale of organic products in the state under their logo, but sales undoubtedly occur without such verification. For the most part, control currently remains more in the hands of the private inspection bodies and their own equivalence checking mechanisms.

2.6   Harmonisation

As is evident from the above discussion, a common understanding of (1) what constitutes organic production, (2) standard guidelines for the operation of organic inspection bodies and (3) an understanding of equivalence, is crucial to the further development of organic production and the organic market. This has been recognised for some time and has been a prominent objective of IFOAM since its inception.

2.6.1 *International Federation of Organic Agriculture Movements (IFOAM)*

IFOAM is a non-profit organisation which brings together the wide range of organic movements around the world and is active in exchanging knowledge, representation, standards development, and in developing an international guarantee of organic integrity. IFOAM Basic Standards were the first internationally agreed reference. The European Commission drew heavily on the Basic Standards in preparing the EU Regulation. In turn the Codex Alimentarius Commission (see section 2.6.2) has referred to the EU Regulation in developing its guidelines. In this way IFOAM has had a major harmonising effect on organic standards worldwide.

In 1992 IFOAM established the IFOAM Accreditation Programme to provide a means of harmonising standards and certification worldwide. The programme offered independent evaluation of inspection bodies against the Basic Standards and the developed IFOAM criteria of organic certification programmes. In 1997 they licensed the International Organic Accreditation Service to perform this function.

International Organic Accreditation Service (IOAS)

IOAS was established by IFOAM as an independent non-profit organisation and operates now from North Dakota, USA. At present fourteen certification programmes are IFOAM-accredited, with six other programmes undergoing evaluation. The application and evaluation process can take several years, involving document review and an on-site visit. The aim of the programme is to provide an independent quality guarantee for organic products and a basis on which certification programmes can accept the products from each other's operators. The IFOAM accredited programmes have recently (October 1999) signed a multilateral agreement to this effect. An IFOAM seal, only to be used on products certified by an IFOAM accredited programme, was also launched in 1999.

Many more certification programmes exist outside of the IFOAM Accreditation Programme than are inside, mainly for cost, technical and political reasons. Part of the problem for IFOAM is that, despite the general acceptance that they represent the worldwide movement and that IFOAM accreditation is respected as a rigorous evaluation, it is only formally recognised by a few individual governments. It is generally understood that, at least informally, IFOAM accreditation has been accepted by most EU member state competent authorities as sufficient proof of equivalence. IFOAM accreditation is not, however, recognised as a mechanism for determining equivalence within the EU Regulation Article 11, despite heavy lobbying by IFOAM. Its accreditation criteria are a sector-specific adaptation of ISO65, which IFOAM believes to be a more appropriate guideline for operating organic certification programmes than EN4501, as required in the Regulation. The Commission's stance is that IOAS must gain membership of the International Accreditation Forum before IFOAM accreditation can be recognised formally. The negotiation continues.

2.6.2 *Codex Alimentarius (Codex)*

Codex is an inter-governmental body, which oversees the joint FAO/WHO Food Standards Programme, which sets international standards aimed at facilitating international trade while protecting consumers from deceptive and fraudulent practices. Codex standards are important in the context of trade rules and conflicts, which may be referred to the World Trade Organisation. Under the WTO Agreement signed in 1994, member countries are encouraged to ensure that standards are not prepared, adopted or implemented in a way that indirectly creates unnecessary obstacles to trade. Governments are required to ensure that private standard-setting bodies also meet this objective. What impact these guidelines will have on organic standards and certification remains to be seen.

Two Codex Committees impact upon the organic sector: the Committee on Food Labelling, working on guidelines for the production, processing, labelling and marketing of organically produced foods, and the Committee on Food Import and Export Inspection and Certification Systems, working on guidelines for organic inspection and certification systems in international trade.

Various groups including the EU and IFOAM are lobbying hard to ensure conformity with their current standards and procedures. The Food Labelling Committee has reached agreement on all areas other than the animal husbandry guidelines and criteria for inclusion of new substances.

2.7   Future developments

New guidelines, standards, amendments and legislation regulating organic operators appear to be emerging daily as we enter a new millennium. Some of the major issues to emerge in the next few years include the following.

Europe

Amendments to Annexes II and VI.
New mechanisms for closer monitoring of imports.
New mechanisms for granting authorisation to imports.
Implementation of the animal husbandry rules.
Rules on labelling of animal feeds.
An overall review of the EU Regulation.

USA

Publication of the second draft of the Organic Rule.

Japan

Publication of the first draft of the Organic Law.

Harmonisation

Publication and adoption of the Codex guidelines.
A new understanding of equivalence.
The role of IFOAM accreditation.
The role of Codex and WTO.

References

Balzer-Graf, U. & Balzer, F.M. (1988) Steigbild und Kupferkristallisation – Spiegel der Vitalaktivität von Lebensmitteln. In: *Lebensmittelqualität – ganzheitliche Methoden und Konzepte.* Alternative Konzepte 66. (eds A. Meier-Ploeger, H.Vogtmann), pp. 1163–1210. C.F. Muller Verlag, Karlsruhe.

European Commission (1991) *Official Journal* No. L198, 22.7.1991, S.1.

European Commission (1999) *Official Journal* No. L222/1, 24.8.

Soil Association (1967) *Mother Earth*, **14** (8), 537.

3 Organic Certification and the Importation of Organically Produced Foods

John Dalby with Michael Michaud and Mark Redman

3.1 Introduction

Any farmer, grower, or food and drink company considering entry into the organic sector faces a bewildering array of considerations. Apart from the practical and financial implications, any business that is serious about going 'organic' will eventually have to consider certification. The purpose of this chapter is to explain the principles and practices of organic certification, with emphasis on the European Union (EU) situation. The chapter also considers European certification within an international context. US certification is reviewed in Chapter 2.

3.2 Organic certification and its importance

Organic certification is generally acknowledged as playing a vital role in the production and marketing of organic food. Initially, it may appear a somewhat bureaucratic burden for the otherwise 'grass-roots' organic movement, but certification remains the cornerstone of a healthy organic market. This is especially so in our modern, industrialised society where consumers have become increasingly separated from food production by a long and often complex processing, distribution and marketing chain. A consumer's choice to buy organic food or drink from a modern retail outlet must therefore be founded upon the knowledge and confidence that the products on sale are truly organic. Consumer confidence can be sustained by an organised system of inspection and certification. It is this inspection/certification system that separates organic products from the plethora of 'health', 'green' and 'ethical' foodstuffs that are frequently grouped together in consumers' minds. Organic certification was therefore developed, firstly to provide an identifiable label for organic food (usually a symbol or logo), and secondly to assure consumers that foods bearing such a label were truly organic throughout the journey 'from plough to plate'.

In order for the certification process to effectively back up the claim that food is organically produced it must involve three principles (Blake 1990):

(1) The setting of organic production and processing standards
(2) Verification that these standards are being followed
(3) Approval of producer/processors and the issue of an organic licence permitting the use of the organic label on specified products.

Therefore, if a farmer or food manufacturer holds an organic production licence it shows that an independent organisation has visited their farm or factory, inspected their production/processing practices, and is satisfied that they comply with an established set of documented organic standards. This is then conveyed to customers of the business via the use of an organic 'label'.

Such assurance is essential, not only in developing and retaining consumer confidence in organic food (especially where it is seen to be more expensive than equivalent conventionally produced food), but also in maintaining the confidence of everyone involved in the trading of organic food whether they be local wholesalers or the multiple retail giants. By deterring unscrupulous 'opportunists', organic certification underpins the high ethical standards of the organic movement and contributes to the encouragement and support of genuine organic farmers and growers. This is especially important in those countries where organic producers are solely dependent upon premium prices in the marketplace, rather than government-funded aid schemes, for providing financial compensation for the extra costs they have incurred through farming organically.

Inevitably, because of the assurance it provides, organic certification plays an essential role in the 'branding' of organic products, an issue of major interest to processors and retailers with an eye on the market for 'green' and 'ethical' foods. In some cases, this branding may relate to a specific symbol or logo. A survey in the UK showed that 85% of people buying organic food looked for an organic symbol before purchase, and that 96% of these were most familiar with one symbol, that of the Soil Association (OFFC 1992). However, with the increasing internationalisation of trade in organic products, consumers are becoming more aware of other European and worldwide labels and are starting to accept these logos and symbols or the terms 'organic', 'biological' or 'ecological' as being a sufficient guarantee of the integrity of the product.

3.3   Organic standards and certification protocol

Organic standards are the detailed rules defining firstly, the production and processing practices that are permitted in the growing and manufacturing of organic food, and secondly, the precautions that must be taken to protect the integrity of an organic product or process.

Organic standards were pioneered by the Soil Association in the UK, who introduced the world's first organic certification scheme in 1973. Basic standards on organic agriculture and food processing have also been laid down by the International Federation of Organic Agricultural Movements (IFOAM 1992a) and are intended as a baseline from which other organisations can develop their own standards.

IFOAM, founded in 1972, is an international, non-profit-making federation representing organisations involved in organic production, certification, research, education and promotion. It began evaluating the standards and inspection/certification procedures of member organisations in 1986, and has published an *Accreditation Programme Operating Manual* (IFOAM 1992b). It has defined three principal requirements which a certification organisation must possess in order to adequately conduct inspections and certifications:

(1)   Competence – the organisation must be financially sound and have sufficient resources and qualified personnel to operate competently.
(2)   Independence – the organisation must operate without interference from vested interests.

(3) Transparency – standards, procedural records and lists of certified operators must be available to interested parties (confidential documents can be excluded).

The manual also includes a detailed description of the structure and operating procedures of the programme and contains the criteria against which a certification programme is assessed.

3.4   Development of organic certification as a legal requirement

Until recently the certification of organic products was mainly a voluntary activity. It carried relatively little legal status and was thus inevitably subject to some laxity. This has changed as organic certification becomes increasingly subject to a range of legislative mechanisms. This legislation can operate nationally and internationally.

There are now agreed guidelines for regulating organic trade under the Codex Alimentarius, an organisation established by the Food and Agriculture Organization (FAO) and the World Health Organization (WHO) to define guidelines for global food standards.

In the US, for example, a number of states have laws regulating organic production and processing. While the usual model is for inspection/certification to be carried out by private sector bodies, these activities are sometimes carried out by state agencies. The US has also been going through the process of introducing a national organic law with the Organic Foods Production Act approved by the US Congress in 1990 (US Senate 1990).

Within the European Union (EU), the principles of organic certification became law on 1 January 1993 when it became illegal to market a wide range of foods intended for human consumption using the term 'organic' in the UK or 'biological' and 'ecological' in other member states unless they fulfil the requirements of Regulation (EEC) 2092/91 (hereafter referred to as 'the Regulation') on organic food production. This is a food labelling regulation and requires that the products must:

(1) Satisfy fully the production, processing and labelling rules contained within the articles and annexes of the Regulation.
(2) Have been properly inspected and certified by an approved body during production and/or processing.
(3) Have been imported from third countries (outside the EU) where the production and inspection procedures are equivalent to those of the EU and have been subject to evaluation either by the European Commission or by a member state.

This means, for example, that it is no longer simply enough for farmers to grow crops organically. If they want to sell their crops as organic, the Regulation requires inspection and certification as proof of their compliance with organic standards. Incidentally, the sale of organic food is defined as 'any supply of products for human consumption in the course of business, including possession, offer and exposure for sale'.

If farmers choose not to register for inspection, but still continue to sell their produce as organic, then in the eyes of the law they are acting fraudulently and could risk prosecution.

In June 1999 the EU Council of Ministers adopted an amendment to the Regulation which includes the additional standards for the inspection and production of organic livestock. The amendment will become law in August 2000 and from then the Regulation will cover products containing both plant and animal ingredients.

3.5   The structure of organic certification schemes in the EU

Since the introduction of the Regulation, compulsory schemes for the certification of organic products have been established across the EU. As the basis for certification, the Regulation establishes the rules for production, processing, and importing, including requirements for inspection, record-keeping, labelling and marketing. It also charges each of the 15 member states to appoint their own control body or competent authority to oversee implementation of the Regulation. Since only the minimum requirements for certification are defined, the national control bodies have a certain degree of flexibility in their interpretation of the Regulation.

This flexibility in the organisation of the certification was required to cater for the existing divergent systems between member states when the Regulation was drawn up. It has resulted in different interpretations of the Regulation in each of the member states (Irish Federation of Organic Associations 1993), and has given rise to a complexity that can be difficult to understand. To help clarify the situation, differences between three sample countries, the United Kingdom, Germany and Denmark, are given below.

United Kingdom

In the UK, the competent authority is the Ministry of Agriculture, which delegates the control of the organic industry to one national control body known as the United Kingdom Registry of Organic Food Standards (UKROFS). UKROFS is made up of representatives of the industry appointed by the Minister. In its capacity as the control body, UKROFS has issued its own standards which include permitted production inputs, processing aids and additives, and appropriate conversion periods as prescribed by the Regulation. In addition, the UKROFS standards have incorporated further rules not addressed by the Regulation, such as those for livestock production and environmental protection.

The Regulation also charges each national control body to register and oversee organic certification bodies where these exist. UKROFS has registered six private certification bodies, which it evaluates annually for their ability to competently certify organic operations (Table 3.1). In addition, UKROFS itself can act as a certification body, for any person or organisation which does not wish to be registered with a private body, although this is not a preferred option. Each certification body must incorporate the UKROFS standards into its own but can add its own rules in so far as they do not violate those of UKROFS. The Soil Association Standards for Organic food and Farming for example, have additional rules on animal welfare and put further restrictions on the use of certain inputs, while the Biodynamic Agriculture Association has included rules relating to biodynamic farming.

Germany

Each *Lander* or region in Germany has its own control body responsible for implementing the Regulation. As in the UK, the private inspection bodies must register in each *Lander* where they wish to conduct business. Since there are a large number of certification bodies that are operating in more than one *Lander* the situation can be cumbersome.

Denmark

Contrasted with Germany and the UK, the situation in Denmark is relatively simple. In this case the control body, a government agency, is the only organisation permitted to certify. There are, consequently, no private certification bodies.

The Regulation requires that the control bodies in each member state allocate each of the certification bodies a unique code which must appear on the product labelling of all the operators registered with them. This is to permit any interested party anywhere in the world to trace the certification body responsible for the certification of the product, and the code should always be applied by an operator to the organic products leaving their premises. In the UK, UKROFS has designated the codes 'Organic Certification UK1 to UK9' to the certification bodies under its jurisdiction.

In addition, from 1 January 1998, every certification body operating within the EU, and also certifying organic products in third countries exported to the EU, must demonstrate

Table 3.1 UKROFS-approved organic certification bodies in the UK and their identification codes.

Identification code	Certification body	Contact details
Organic Certification UK1	The United Kingdom Register of Organic Food Standards (UKROFS)	c/o MAFF, Room 114, Nobel House, 17 Smith Square, London SW1P 3JR. Tel. 0207 238 6004. Fax 0207 238 6148. UKROFS can also inspect individual producers and processors, but this is not a preferred option.
Organic Certification UK2	Organic Farmers & Growers Ltd (OF&G)	Churchgate House, 50 High Street, Soham, Ely, Cambridgeshire CB7 5HF. Tel. 01353 720250. Fax 01353 720289.
Organic Certification UK3	Scottish Organic Producers Association (SOPA)	c/o Milton of Cambus Farm, Doune, Perthshire FK16 6HG. Tel. 01786 841657. Fax 01786 842264.
Organic Certification UK4	The Organic Food Federation (OFF)	Unit 1, Manor Enterprise Centre, Mowles Manor, Etling Green, Dereham, Norfolk NR20 3EZ. Tel. 01362 637314. Fax 01362 637398.
Organic Certification UK5	Soil Association Certification Ltd (SACert)	Bristol House, 40–56 Victoria Street, Bristol BS1 6BY. Tel. 0117 9142 400. Fax 0117 925 2504.
Organic Certification UK6	Biodynamic Agricultural Association (BDAA)	Painswick Inn Project, Gloucester Street, Stroud, Gloucestershire GL5 1QG. Tel. 01453 759501. Fax 01453 759501.
Organic Certification UK7	Irish Organic Farmers & Growers Association (IOFGA)	Harbour Building, Harbour Road, Kilbeggan, County Westmeath, Ireland. Tel. 00 353 506 32563. Fax 00 353 506 32063. NB Generally operates in Northern Ireland.
Organic Certification UK number not allocated	Food Certification (Scotland) Ltd (Organic Certification of Farmed Salmon in the UK)	Redwood, 19 Culduthel Road, Inverness IV2 4AA. Tel. 01463 222251. Fax 01463 711408.
Organic Certification UK9	Organic Trust Ltd	Vernon House, 2 Vernon Avenue, Clontarf, Dublin 3. Tel./fax 00 353 185 30271.

that the criteria of EN45011, or its international equivalent ISO65, are being complied with. These standards have been internationally recognised for bodies operating inspection and certification of products and specify the criteria by which these bodies must operate. While it is not necessary for the bodies to be accredited with the national EN45011/ISO65 accreditation body, many have done so and, at the time of writing, most UK certification bodies have made an application to the United Kingdom Accreditation Service (UKAS) for full accreditation. As an interim measure, the UK bodies have been audited by MAFF for UKROFS in order to demonstrate that their operating procedures are equivalent to the EN45011 criteria.

The Regulation and the national or private standards, developed by control and certification bodies, should not be considered to be static documents. Changes can be made to respond to the needs of the industry, and there are avenues by which changes can be initiated. The principal means of amending the Regulation is through the Article 14 Committee, which was established by Article 14 of the Regulation. This committee is made up of government representatives from each member state and receives proposals from the states, made through the certification bodies, to add or remove permitted materials to the annexes or amend the text. Where a consensus is reached, the Article 14 Committee makes recommendations to the EU Commission and eventually the proposals may become integrated into the Regulation. The Commission has also taken the unprecedented step of asking expert witnesses from private certification bodies in some member states to attend meetings in Brussels to discuss particular problems.

In addition, there is an IFOAM group, which advises the EU Commission on matters pertaining to organic agriculture. This group is made up of representatives from each of the IFOAM members in the EU, and provides a non-governmental route for advising at an EU level.

3.6   Certification protocol

Despite the differences in the structure in organic certification between EU countries, the basic protocol remains the same regardless of which certification body is involved. This protocol (Table 3.2) involves the following steps:

(1)   Determination of eligibility by the operator, i.e. does the operation comply with the production and/or processing standards?
(2)   Completion of the requisite application documents provided by the certification body.

Table 3.2   Certification protocol.

Stage	Action
Determination of eligibility	Undertaken by the applicant or consultant
Application	Undertaken by the applicant or consultant
Inspection	Inspector employed by the certification body
Certification appraisal	Certification officer or committee
Decision	
Approval	Certificate issued
Conditional approval	Compliance notice issued
Failure	Notice of failure issued

(3) Inspection visit to the applicant's holding or processing premises.
(4) Evaluation of the information in the inspection report, followed by the certification decision.

3.6.1 *Determination of eligibility of a business for organic certification*

Having decided the need to be certified, the producer or processor of organically produced foods must then determine his/her eligibility for certification. Eligibility means compliance with the Regulation (where applicable) as well as any further rules imposed by the certification body chosen by the applicant. Standards for organic production and processing are available from the relevant bodies, and the applicant should obtain a copy to review. Table 3.3 summarises some of the main points of the Regulation relative to the production and processing of organic foods.

Agricultural and horticultural crop production

One of the primary concerns of organic production is the development of biologically active, sustainable systems that minimise any negative impacts on the environment. Some of the main features are therefore as follows:

*   Limited use of inputs derived from outside the system. Instead cropping involves the inclusion of legume crops and green manures, the cultivation of adapted species and varieties and the formulation of appropriate rotation strategies.
    Approved inputs are listed in Annex II of the Regulation although nearly all require permission from the certifying body before they can be used. Approved fertilisers and soil conditioners include low soluble materials such as rock phosphate and lime, as well as animal manures and by-products such as bone meal and hoof/horn meal. More soluble materials such as blood meal, wood ash and potassium sulphate are also allowed, again following approval from the certification body.
*   Insect control is restricted to natural insecticides derived from plant extracts, such as pyrethrum or derris. Biological controls such as *Bacillus thuringiensis* and predator

Table 3.3 Determination of eligibility for organic certification.

Operators	Requirements that must be demonstrated to the certification body
Producers	Use of approved inputs Compliance with appropriate conversion periods Use of practices which maintain or increase biological activity during and after conversion Avoidance of parallel cropping of organic and non-organic crops on the same unit Maintenance of adequate records Compliance with storage, transport and packing standards
Processors	Use of organically grown products Compliance with standards regarding additions of non-organic ingredients, additives and processing aids Separation of organic and non-organic processing and storage Prevention of external contamination Identification of process lots and maintenance of adequate records Compliance with storage, transport and packing standards

insects and mites are also allowed. Permitted materials for disease control include copper-based fungicides and sulphur. Although approved for use in organic farming, national laws may restrict the use of these materials to licensed products.

- Materials not allowed include highly soluble mineral fertilisers, whether natural or synthetic, such as potassium chloride, urea, Chilean nitrate, single and triple superphosphate and synthetic insecticides, fungicides and all herbicides.

For a crop to be labelled as 'organically produced', a conversion period of two years from the last use of non-permitted inputs to the planting of an annual crop and three years to the harvest of a perennial crop is required. During this period only approved inputs can be used. In addition to the 'organically produced' category, the sale of in-conversion produce was allowed until 1 July 1994. For a crop to be sold under the official label 'product from land in conversion to organic farming', the crop must be harvested from land which has been registered as in-conversion for at least 12 months.

There is no requirement under the regulation for an entire agricultural holding to be fully converted to organic agriculture. However, in those cases where crops are grown both organically and non-organically, parallel cropping is not permitted. This means that plants of the same variety as those produced on the organic unit cannot be produced on the non-organic unit. During harvesting, storage and transportation, organically grown crops must be protected from contamination by residues of non-organically grown crops and chemicals used in cleaning, fumigation and pest control. The crops must also be handled and identified in such a way as to prevent any confusion between organic, non-organic and in-conversion grades. In addition, storage areas should be covered to prevent contamination by bird droppings and protected from vermin such as rats and mice.

Accurate and up-to-date records must be kept of the production activities and must be of a nature to demonstrate to the inspector that the standards of the certification body have been met. In the case of a mixed unit where crops are grown both organically and non-organically, records must be kept for both units. They include origin, nature and quantities of bought-in materials such as quantities of each product sold.

The UKROFS Standards specify that cropping records must also be maintained on a field-by-field or area-by-area basis and must include the following:

(1) The cropping history, including crops and yields.
(2) The rate, type and date-of-use of products employed for fertilisation, soil conditioning and weed, insect and disease control. For land in conversion, these records should be maintained for previous treatments over the last three crop years.
(3) The source and type of seeds and/or transplants used (including any chemical treatments during propagation). This is particularly important in cases of potential parallel cropping.

Animal products

Until the implementation of the amendment to the Regulation in 2000, livestock and animal products are not covered. However, the national control bodies and organic certification bodies have formulated standards to include them and these will have to be modified to conform to the new Regulation by the summer of 2000. An additional article will permit the control bodies in member states to set more stringent standards where they wish while still having to accept as organic animal products produced by all the other member states. The Regulation will become law on 24 August 2000. By that date UKROFS will have decided which sections of the Regulation the UK will retain in a more restrictive form.

While the current livestock standards may vary from control body to control body and even between the private certification bodies within a member state, the main issues remain the same.

Welfare. Animal welfare considerations are a key aspect in organic production methods. Generally speaking the animal must be allowed free movement, access to the outside and association with members of its own species. Handling and transport of live animals and birds must also consider their welfare.

Source of animals. While a conventional herd or flock can be converted to organic production, there are usually limitations on the numbers of animals that can be brought in from non-organic sources as replacement animals within a particular year. For those conventional animals that are converted or brought-in, there are conversion periods before the milk and offspring can be considered as organic. In the UK, animals for meat production must be born on a registered unit.

Feeding. The main component of the diet should be organic and in-conversion forage, but often the use of small percentages of non-organic feed is allowed. These percentages can vary among the classes of livestock, with ruminants allowed the least and monogastrics the most. Types of feed may also be restricted. For example, various plant-based meals from which oils have been solvent-extracted are prohibited, as are all feedstuffs containing genetically modified ingredients.

Veterinary medication, sprays and dips. Generally speaking, routine prophylactic treatments are not allowed but animals can be treated with the appropriate veterinary medicines when ill or injured. Withdrawal periods are longer than the statutory ones, and more appropriate management of the stock, such as reduction of the stocking rate, might be necessary to limit dependence on medicine. In the UK, at least, organophosphate dips and treatments are not permitted under any circumstances.

Slaughtering and meat processing. As with animal production, the original Regulation did not cover slaughtering and meat processing though, in the UK, the UKROFS standards specify that abattoirs, cutting plants and butchers should be certified if the meat is to be sold as organic. These requirements have been incorporated into the new amendment and will become law in 2000.

Record-keeping. Animal production records must be kept as proof of conformity to organic standards. These include movements of animals entering and leaving the farm, veterinary treatments, feedstuffs and feeding regimes.

Processing and packing

Any operator processing, manufacturing or packing a food product composed wholly or partly of organic ingredients and marketed in such as way that there is a reference to organic production must be registered with an approved certification body. These operators include the following:

(1) The 'typical' food processors such as those making ice cream, cheeses, cider and wine.
(2) Traders, retailers, and wholesalers, including packers and prepackers of grains, fruits and vegetables, who break down, relabel or repack bulk material 'out of sight of the final customer'; this category includes health-food shops who buy in large sacks of organic produce and repack them into smaller retail packs.
(3) Wholesalers who buy in organic products in bulk for storage on their premises with the intention of reselling them on; this category includes grain traders.
(4) On-farm processors such as dairy farmers making farmhouse yogurt and cheese using their own milk.

Product types and labelling

The Regulation sets out the rules of processing, including labelling and product identification. As a labelling regulation it gives a legal definition to the terms 'organic', 'biological' and 'ecological' when applied to foods for human consumption in the various member states. Ingredients other than organically produced agricultural products are allowed, though types and quantities are controlled. These ingredients are listed in Annex VI of the Regulation, and any operator contemplating the processing of organic foods should refer to it.

There are only two categories of foods which can carry indications of organic status, based on the proportions of organically produced ingredients:

(1) A product can only be described as 'organic or 'organically produced' if not less than 95% of agricultural ingredients have been organically produced.
(2) A product can carry the description 'Made with X% "organic" or "organically produced" ingredients' if not less than 70% of the agriculture ingredients have been organically produced.

Products not falling into either of these categories must not carry any indications of the organic status of the ingredients, whether in the product name, the ingredients panel or on the sales literature.

In both cases, all the other ingredients can only come from positive lists given in the Annexes VIA, VIB, and VIC such that:

(1) The remainder of the agricultural ingredients can be non-organic, provided that they appear on the limited list given in Annex VIC.
(2) Any processing aids used in the production, such as releasing oils and flushing gases, and so on, must be listed in Annex VIB.
(3) Any ingredients of non-agricultural origin, such as additives, yeasts, minerals, and

so on, must be listed in Annex VIA, though the quantities used in the production of organic food are not limited.

(4)   The ingredients must not have been subject to treatments by ionising radiation or be derived from genetically engineered plants and products.

Where any approved non-organic ingredients are present, they should be differentiated from organically produced ingredients in the ingredients panel. This is often done by means of an asterisk against the organic ingredients with the definition – *organically produced ingredients – elsewhere in the panel. All ingredients must also be listed in descending order with the individual weights or percentages of the total weight, as weighed into the mix, against each one. Processing aids are not listed in the ingredients panel since, theoretically, they are not constituents of the final product.

Where the labelling or advertising materials relating to a product carry an indication of 'organic', this indication must refer to a method of agricultural production, such as 'organically produced' or 'product from organic farming' The term 'organic flour' may be unsafe in a legal sense and 'flour made from organically produced wheat' may be the more appropriate form of wording. However, the common usage of the term 'organic' has been accepted provided that elsewhere on the label or packaging this is supplemented with a reference to a farming practice. If in doubt, the local officials dealing with trade descriptions should be consulted in this situation, for example, the Trading Standards Office in the UK.

Processing equipment and operations

The processing of organic products must be done in such a manner as to prevent contamination or accidental substitution of organic and non-organic food products. All processing operations must be registered with the local health officials (Environmental Health Office in UK) and comply with the relevant food safety regulations.

Ideally, the processing of organic foods should take place in separate and dedicated sites or, failing that, using separate and dedicated equipment. However, this is not always possible and, within a non-dedicated context, separation can be achieved in time by batched production. A specific time is allocated for the processing of the organic food, whether it is a particular day or a certain time of the day. If done during a certain time of the day, the first run is preferable for organic foods in order to take advantage of the previous day's cleandown, thus minimising contamination from non-organic food residues. The operations must be carried out continuously until the organic production run has been completed.

Equipment used for processing should be made from non-porous food-grade materials and be subject to the appropriate cleaning operations. A rinse with potable water should precede the organic production run to remove any residues of cleaning chemicals. If it is not easily dismantled for manual cleaning or inappropriate for cleaning in place (CIP) with liquid cleaners, the equipment must be subject to a bleed run with the organic product to purge the system of non-organic residues.

Storage/warehousing

Incoming raw ingredients of an organic nature should have dedicated storage areas.

These areas should be appropriately identified and separated from those containing non-organic ingredients by sufficient space or physical barriers to prevent confusion and cross-contamination. Birds, insects and vermin should not have access to these areas. Work in progress and finished goods should also be stored in clearly labelled and designated areas.

Transport

Transportation must be done in a manner to prevent the contamination and accidental substitution of the food product being moved so as to maintain its organic integrity. Consequently, the vehicles and containers involved should be subject to a regular cleaning and inspection programme to prevent the build-up of non-organic residues. For the transport of bulk or wholesale loads between a registered and non-registered unit, the Regulation requires that the sacks or boxes be sealed. This restriction on sealing, however, does not apply to the transport between two registered units. In both cases, the wholesale packs must be provided with a label and/or document which states the name of the processor/packer, name of the product, organic nature of the product, and the certification body responsible for certifying the producer/processor. An example of the above is the bulk transport of grain from a registered grain store to a registered flour mill. Since transport is done in a lorry which is not practical to label or seal, a delivery note or invoice should accompany the driver to appropriately identify the load. The load is, of course, covered to prevent contamination, while the trailer must be cleaned prior to being loaded. If the grain is placed in sacks before transportation, each sack would have to be individually labelled.

Record-keeping

Both producers and processors must keep accurate records of their activities. Records must be kept in such a manner that the organic raw material used in the finished product can be traced back to the original source, and that a reconciliation of input versus output can be done. Records should also include a recipe or product specification sheet which list the ingredients and their weights. The dates and quantities processed or packed must also be kept, and a batch number or use-by-date must be put on the packaging or container to allow trace-back to the processing day. Delivery notes and sales invoices must also be kept. In addition to processing records, cleaning and pest control records must be kept. These include cleaning schedules and a list of all substances used and details of any fumigation treatments such as dates of treatment, chemicals used and commencement of processing.

3.6.2 *Making an application*

After determining the eligibility of the operation, the interested party must request an application form from the certification body. The applicant completes and returns it to the certification body, where it is reviewed for completeness.

As part of the application process, farmers and growers must provide a description of both their organic and non-organic units. This description should include land areas, storage areas, and any on-farm processing and packaging which may take place. Maps must be provided and field histories, including the last application of prohibited inputs on the organic fields, must be provided.

As part of their application procedure, processors, packers and distributors must supply recipes or product specification sheets outlining all the ingredients and their percentages by weight. They must also indicate if the unit is dedicated and supply a description of the unit, including plant and equipment, and warehousing and storage facilities. The processing operation, must be described (including a flow diagram); cleaning operations and pest control procedures are also included, as are documentation and record-keeping protocols.

For operators that have already been certified, new applications do not have to be made annually for their existing enterprises. However, as new products, fields or animals are brought into the scheme, an application will have to be made.

3.6.3 *Inspection*

Preliminaries

After the initial screening of a new application by the certification body, the inspection can take place. The regulation requires that inspections take place annually, though operations often may be inspected more than once a year. This is particularly true where the operations are somewhat complex and the whole unit cannot be seen during one visit. Usually, one inspector at a time will visit a unit, though occasionally it is appropriate for more than one to conduct the inspection. Inspectors are chosen by the certification body to inspect a unit based on their expertise and geographical proximity to the site. Operators, however, can refuse an inspector when they feel the inspector for one reason or another may not be objective. Then, too, the inspector may refuse to do a job, perhaps due to some conflict of interest.

Inspection protocol

Inspections can be facilitated by proper preparation on the part of the operator. Tables 3.4 and 3.5 list the information and records required at the inspection. It is useful for the operator to be well-organised and thus contribute to a smooth running and trouble-free inspection.

The protocol to be followed is outlined in Table 3.6. One of the main objectives of the inspections is to check conformity with the standards set by the certification body. In the case of reinspections the purpose is also to check compliance with the conditions set by the certification body as the result of a prior inspection.

The inspections are a combination of information gathering and information verification. While the application form and annual return, which are filled in by the operator, contain a great deal of information, it is the inspectors' job to verify this information. They also collect information not normally supplied in the annual returns and applications, such as potential spray drift, constituents of animal feeds and dates of fumigation and processing.

Within the inspection structure of information collection and verification there is an inspection methodology consisting of four activities. These activities and an explanation of each are as follows:

Table 3.4 Information, records and paperwork required at producer (from Soil Association Certification Ltd).

Category	Information
General	Field histories
	Crop
	Inputs, e.g. fertilisers, farm yard manure, pesticides
	Crop and forage production
	Rotations
	Manure and fertiliser management, including sources, treatments, rates and dates of applications
	Seed treatments
	Pest and disease-control measures
Horticulture	Module numbers and source
	Composts
Livestock	Animal numbers, breeds and ages
	Feed items of both home-produced and bought-in materials
	DM content
	Ingredients
	Stock movement record
	Bought-in
	Sold
	Born
	Stock identification
	Veterinary records
	Housing area for each class of livestock
Financial records	Purchase invoices
	Sales receipts
	Accounts

(1) *Interviews*. This consists of talking with the operator, those in charge of managing the unit, and other employees. The purpose of these interviews is to ascertain personnel awareness of production and processing, as well as test their knowledge of the standards. They may also give further background on the operation and provide more information where necessary.

(2) *Farm/factory walk*. During this activity the inspector checks for spray drift and chemical storage areas on farm. They may also weigh out feed to determine the accuracy of declared feeding regimes, while checking veterinary storage areas for any undeclared use of medicines. In processing units they might check storage areas or the operating procedures for the separation of organic and non-organic products.

(3) *Records check*. This is becoming an increasingly important activity. In factories the inspector may carry out a reconciliation between raw materials and finished goods to see if the figures match. They may check fumigation records such as the use of methyl bromide and the date when organic processing began in order to determine potential contamination. In a production unit they may check purchase invoices to determine seed treatments and the purchases of chemical fertilisers to detect if any of these have leaked onto the organic unit.

(4) *Soil or product sampling*. In cases of suspected contamination, the Regulation requires sampling of the product to check for chemical residues. The sampling may be done on

Table 3.5 Information, records and paperwork required at processing inspections (from Soil Association Certification Ltd).

Category	Documentation
Certification documents for imported products	The originals of the EU certificates which accompany consignments of organic produce from countries outside the EU
	Letters from the control body authorising the importing of products from outside the EU
	Certificates of certification accompanying consignments of organic produce
Documentation for goods received	Delivery notes and invoices
	Goods received logs and/or records
	Records confirming the authenticity of the organic goods
	Certificates of certification accompanying consignments of organic produce
Production records	Processing records and production logs
	Product specification sheets for existing and new products
	Bleed runs used to purge equipment which cannot be cleaned before use
	Best-before dates or batch numbering systems
Sales records	Totals of organic products sold
	Sales invoices and delivery notes
Stock taking records	Stock taking records
Hygiene/cleaning schedules	Hygiene/cleaning schedules before and during organic production runs
Pest control records	Materials used by pest control contractor
	The dates of the applications of pest control materials

Table 3.6 Inspection protocol.

Stage	Description
Inspection objective	To check conformity with the Standards
	To check for compliance with the conditions set by certification body
Inspection structure	The verification of the information supplied with the application
	The collection of additional relevant information
Inspection methodology	Interview with the responsible person
	Checking the records and documentation
	Soil/product sampling
	Physically inspecting the farm or the factory
Inspection debriefing	Explaining any areas of uncertainty in the Standards
	Discussing any problems or irregularities identified with the responsible person and getting their agreement to correct these
	Getting the inspection report signed by the responsible person
Post-inspection	Writing the report and making the recommendations for the certification body to act on

raw ingredients in store, in fields where drift has been suspected or in processing units where inadequate bleed runs or cleandowns may not have been practised.

As the inspection progresses, the inspector completes a questionnaire which covers all the points relevant to the certification of the operation. At the completion of the inspection, the operator is required to sign the questionnaire, confirming that to the best of their knowledge, all the information contained therein is accurate and correct.

3.6.4 *Post-inspection activities and certification*

After the inspection the inspector writes a report, making the appropriate recommendations concerning the organic production and practices of the unit. The report is then sent to the certification body for consideration. The decision may be taken either by an individual certification officer or a committee acting for the certification body. In general, certification decisions can take one of three routes (see Table 3.7):

(1) *Successful award*. No irregularities are found and certification documents are issued.
(2) *Manifest infringement*. Certification is refused to the entire unit. This would occur, for example, in cases of outright fraud where non-organically produced raw ingredients were being used in processing, or prohibited chemicals were intentionally used for crop production.

Table 3.7   Certification decisions made upon the basis of the inspection report

Inspector's report	Certification decision	Action
Minor non-compliances reported	Conditional approval	Compliance Notice issued which specifies: (a) Additional information to be supplied (b) Immediate correction of minor non-compliance(s) identified Action to be agreed and acted upon by applicant before certificate can be issued
Serious irregularity reported	Certification refused for a product, part of farm or enterprise or lot or production run	Compliance Notice issued identifying the irregularity Applicant agrees to Notice or appeals against decision Control body notified if certification withdrawn from existing operator Additional inspection required to check compliance
Manifest infringement	Certification refused for the entire unit or holding	Compliance Notice issued notifying decision Applicant agrees or appeals against decision Control body notified that certification withdrawn for existing operator
Appeals procedure	Applicant appeals against the decision Appeals procedure deals with an appeal against a decision following further information from applicant	May be upheld or modified Applicant may take the appeal to arbitration if agreement not reached

(3) *Irregularity or minor non-compliance.* This is the most typical result of the certification decision. In this case, certification may be refused at the discretion of the certification body to part of the holding, some of the enterprises or an entire lot/production run. Compliance with the standards is then required, and once this has been demonstrated, the certification can be completed. Compliance may only require the provision of further information such as a map or recipe sheet. It may also involve instant 'correction' such as upgrading some aspects of record-keeping or modifying one or more ingredients in a product. In some instances, the land or animals may have to be submitted to a conversion period. For example, a dairy animal given feeds containing prohibited ingredients may have to go through an appropriate conversion period before its milk can be sold as organic. In the case of land or crop, unless the operator can convince the inspector and certification body that new land being brought into the scheme has been converted according to the standards for the necessary time period, it may have to go through a full conversion period of two years before planting for an annual crop and three years before harvesting a perennial crop.

When violations of standards are found, the certifying body may remove the certification for a production run or batch or for the entire unit. In such cases, the control body must be informed. A fine may also be imposed by additional inspections, or product analysis being specified at the operator's cost. The type of fine will depend on the severity of the violation and may vary from body to body. In the case of a manifest infringement or irregularity, the operator has the right of appeal if they consider the decision to be incorrect. Since certification is decided on information supplied by the inspector, the appeal must be accompanied by relevant information and the reasons why the applicant believes the decision should be changed.

The Regulation requires that all operations seeking certification must give to the certification body a signed agreement stating that they will carry out production and processing in accordance with the standards. In the case of an irregularity, they must agree to remove any reference to organic production from the relevant crop, animal or processing run. Where there is a manifest infringement, the operator must agree to prohibitions on marketing the crops as organically produced for a period of time designated by the certification committee of the certifying body.

3.7　Imports of organically produced foods

Organic foods produced within the EU or imported into one member state through authorised channels must be freely traded and marketed in the other member states without the need for further registration or inspection. Importing from third countries outside the EU is another matter however. The rules vary, depending on whether the food comes from an approved or non-approved third country.

3.7.1　*Imports from approved third countries*

The EU Commission can directly approve countries where a national government has supplied a dossier to the Commission which confirms that the procedures for production, processing, inspection and inspection monitoring by the state, are equivalent to those of the EU.

There are currently only five countries which have been approved by the EU Commission as having equivalent inspection and production arrangements: Argentina, Australia, Hungary, Israel, Switzerland and the Czech Republic. Organic products from these countries can enter the EU if the following conditions are met:

(1) Registration of the importer. This involves the inspection and certification of the importer within the EU by an approved certification body. The inspection covers the offices of the operation as well as any ancillary storage and distributions sites, especially if the handling of bulk materials such as grain is involved. On completion of the certification procedure, the body registers the importer with the national control body, for example, UKROFS in the UK.

(2) Provision of an EU certificate (European Union Certificate for Importing Products from Organic Production; see Fig. 3.1). Each consignment of an organic product must be accompanied by the completed EU certificate as issued by the control body and/or the certification bodies. The original must be supplied by the overseas exporter and be kept on file. A photocopy or fax is not acceptable. Copies may be provided to interested parties such as certification bodies or customers on request, but these should be clearly stamped as being a copy.

3.7.2 *Imports from non-approved third countries*

To obtain approval to import organic products from non-approved third countries is a much more complicated procedure and requires the following steps to be undertaken:

(1) Inspection of the producers, processors and exporter involved in supplying the product. This can be done by an indigenous certification body or one from either the EU or another third country.

(2) Registration of the importer. The procedure has already been described in section 3.7.1.

(3) Application to the national control body. To import the product in question, the importer must make an application to their national control body (in the UK, the application form OB6 must be completed and sent to UKROFS). This application, one of which must be made for each product from each country, must include the following:

 (a) Information about the certification body in the country of origin including a copy of the standards and operating manual. This is to determine whether the certification body's rules of production and inspection are equivalent to those specified in the Regulation.

 (b) A declaration from the certification body that its rules of inspection/certification will be 'permanently and effectively applied'.

 (c) A confirmation that the certification body's operation in the third country satisfies the criteria set under EN45011 or its international equivalent, ISO65. The importer must supply the appropriate documentary evidence which confirms compliance by one of the following three options:

  (i) Full accreditation with the accreditation body responsible for EN45011

EUROPEAN COMMUNITY
CERTIFICATE FOR IMPORT OF PRODUCTS FROM ORGANIC PRODUCTION

1. Body issuing the certificate:	2. Regulation (EEC) 2092/91, Art 11 Reference number of the certificate:
3. Exporter of the product:	4. Control/inspection body:
5. Producer or processor of the product:	6. Country of dispatch:
	8. Country of destination:
7. Consignee of the product:	9. Address of place of destination:
10. Marks and numbers, container number(s) Number and kind: Trade name of the product:	11. Gross weight (kg): 12. Net weight (kg): 13. Alternative units:

14. Declaration of the body issuing the certificate:
This is to certify that the products designated above have been obtained in accordance with the rules of production and on inspection of the organic production method, as set out and monitored by the control organisation mentioned in box 4.

15. Additional declaration (if appropriate):

16. Place and issue of the certificate: Stamp of the issuing organisation:

Name and signature of authorised person: Date:

Explanatory notes:	Box 4	control body for monitoring compliance with the rules on organic production methods;
	Box 5	the firm which carried out the last operation (processing, packaging, labelling) on the batch;
	Box 9	the address of the firm where the batch will be delivered, if different from the address in box 7;
	Box 13	e.g. volume in litres in case of liquids to be given, where appropriate, in supplement to the declarations in boxes 11 & 12

Fig. 3.1 EU certificate (European Union Certificate for Importing Products from Organic Production).

or ISO65. Preferably this accreditation body should be subject to mutual recognition agreements based on peer evaluation put into place by the International Accreditation Forum or, in any EU member state, for the accreditation of certification bodies.

(ii) Confirmation by the control body or government in the third country. The importer must submit the legal basis and the documents by which a control body or government agency in the third country guarantees the conformity of the inspection body with the requirements of EN45011

> or ISO65 and ensures that a periodic surveillance and reassessment is made of the certification body.
>
> (iii)   Confirmation by the control body in the EU member state which is going to grant the authorisation. This must be made by independent and competent experts or official accreditation bodies designated by the control body (other certification bodies cannot do this work).
>
> (d)   The name and address of the agency or individual in the country of origin who will be signing the EU certificate.
>
> (e)   A list of the products and a specification sheet for each if composed of more than one ingredient.
>
> (f)   The names and addresses of the exporter, the processor and the producers in the third country.

(4)   Approval of the importer. Provided that the above is satisfactory, the national control authority then notifies the importer that the import of the product has been authorised. Only after receipt of this authorisation can the importer import and market the product. A copy also goes to the body certifying the importer.

(5)   Provision of the EU certificate. As with the imports from approved countries, a completed EU certificate must accompany each consignment being imported.

3.8   The future development of organic certification

While the basic framework of organic certification – application, inspection, certification – remains static, there is a dynamic element to the process. New additions and deletions are being made to Annex VI, while the permitted chemicals used for insects and disease undergo changes in Annex II. A new livestock regulation has been announced for implementation in 2000. The list of approved third countries will probably be extended, and it is possible that a list of approved third country certification bodies may be drawn up to simplify the procedures and not restrict the international trade in organic goods.

  Because of this inherent dynamism, it is critical that communication channels between operators and their certification bodies remain open. Though certification bodies should inform their operators of any changes in the regulations and standards, it is ultimately the operators' responsibility to be aware of any developments that may affect certification. The worst time to find out is during an inspection.

Acknowledgement

Thanks are extended by the authors to Soil Association Certification Ltd for permission to use information, which has been invaluable in completing this chapter.

References

Blake, F. (1990) Standards: regulating organic food production. In: *Organic Farming – an Option for the Nineties*. British Organic Farmers/Organic Growers Association, Bristol.

Coopers & Lybrand Deloitte (1990) *Going Organic – the Future for Organic Food and Drinks in the UK*. Coopers & Lybrand Deloitte, Birmingham.

FAO (1999) *Guidelines for the Production, Processing, Labelling and Marketing of Organic Foods*. Codex Alimentarius Commission, Food and Agriculture Organisation, Rome.

IFOAM (1992a) *Basic Standards of Organic Agriculture and Food Processing*. International Federation of International Agriculture Movements, Tholey-Theley, Germany.

IFOAM (1992b) *IFOAM M Accreditation Programme Operating Manual*. International Federation of International Agriculture Movements, Tholey-Theley, Germany.

Irish Federation of Organic Associations (1993) Taking Stock of Organic EC Legislation (*Newsletter* No. 6). Irish Federation of Organic Associations, Dublin.

OFFC (1992) *Organic Food – are we getting enough?* A survey report from the Organic Food and Farming Centre, Bristol.

Soil Association (1992) *Standards for Organic Food and Farming*. The Soil Association Organic Marketing Company Ltd, Bristol.

US Senate (1990) *Organic Foods Production Act of 1990*. United States Senate Committee on Agriculture, Nutrition and Forestry, Washington, DC.

4 International Market Growth and Prospects

Carolyn Foster

4.1 Introduction

Over the last two decades the growth of the environmental movement combined with concerns about health and quality of diet have led many people to question modern agricultural practices. Concerns have centred around the safety of food following a number of food scares, such as salmonella, listeria and BSE (bovine spongiform encephalopathy) and, more recently, genetically modified foods, the environmental implications of widespread agrochemical use, as well as the associated health risks of residues in food. The development of the market for organic food has been largely consumer-led in recent years, prompted by the aforementioned issues and concerns and, combined with a period of economic prosperity, the result has been a dramatic increase in demand for organically produced food which is perceived as being healthier and less damaging to the environment. In many European countries, especially in northern and central Europe, demand for organic food has outstripped supply, a situation which still prevails in some countries such as the UK.

4.2 Organic production

Organic agriculture is of growing importance in the agricultural sectors in a number of countries. Increased demand has meant that most organic produce is able to command a price premium, and these attractive prices, coupled with policy support for organic farming, encourage conversion to organic farming. In the European Union, supply has increased dramatically since the late 1980s with land area under organic production increasing from 0.16 million hectares in 1988 to nearly 2.1 million hectares by the end of 1997 (Fig. 4.1), representing 1.5% of total agricultural land area compared with 0.1% in 1988 (Foster & Lampkin 1999). This figure is estimated to have risen to over 2% in 1998.

These figures vary considerably within and between countries. The increase in organic land area has been most notable in the Scandinavian and German-speaking countries. Here, organic farming is moving from a marginal agricultural activity to occupying a significant share of total agricultural land use. Austria now manages over 10% of land area organically, rising to as much as 30% in some regions. In Switzerland, Finland, Sweden and Italy 4–7% of agricultural land is managed organically. Denmark and Germany both have around 2%. In the majority of EU countries, however, organic land area still accounts for less than 1% of total agricultural land.

In Europe, conversion to organic farming was further encouraged when policy makers began to recognise organic farming as a positive way to mitigate problems created by

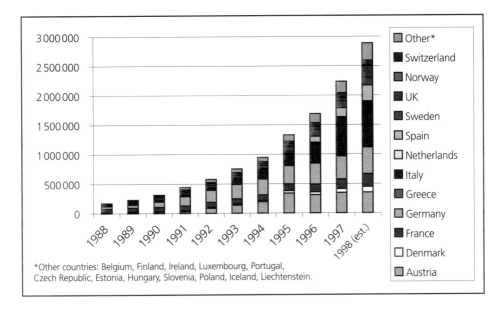

Fig. 4.1 Certified organic and in-conversion land area (ha) 1988-98 [source: Foster & Lampkin (1999, unpublished report) and Lampkin N.H., Welsh Institute of Rural Studies].

the Common Agricultural Policy (CAP). Support policies can be used not only to reduce overproduction and environmental problems, but also to satisfy the market demand for organic products (Lampkin & Padel 1994). In the late 1980s, politicians introduced programmes to support conversion of farms to organic agriculture, in part recognition of the costs incurred by producers during the conversion process and the contribution they make to environmental, animal welfare and social objectives. Attracted by subsidies, farmers in a number of countries, including Austria, Denmark, Germany, Sweden and Switzerland, took advantage of aid schemes for the conversion of land to organic farming between 1987 and 1992. In the UK no measures were introduced until the CAP Reform in 1992.

The introduction of EC Regulation 2078/92 as part of the 'accompanying measures' of the CAP Reform in 1992 provided a framework for the implementation of policies to support organic farming in the EU, whereby farmers wishing to convert to organic farming and existing organic farmers are eligible for aid. Although all member states have introduced such policies for organic farming, levels of uptake between countries vary as do the types of farms converting. There is evidence that this can be related to the level and structure of payment rates, which disadvantage specialist cropping and intensive livestock farms. The impact on the development of the market is yet to be determined (Lampkin *et al.* 1999).

4.2.1 *Market size*

Accurate data for the organic market are scarce and varies according to source. The following sections will attempt to present and summarise data presented in a variety of sources.

Demand for organic food has been growing steadily since the mid-1980s. However, the supply side in many countries has been slow to respond and markets have remained underdeveloped. Fresh impetus came in the early 1990s due to a combination of factors mentioned in the introduction. Although global sales of organic food still account for less than 1% of total retail sales, the organic sector has become one of the fastest expanding areas of the food industry. In the UK, for example, retail sales of organic food have almost tripled between 1992 and 1997, rising from £92m to an estimated £260m (Mintel 1997). Although this represents less than 2% of total food retail sales, the last three years have witnessed growth rates of around 30% per annum and sales are projected to account for a 7–8% share by 2002, reaching a value of over £1bn (Soil Association 1997). Similar growth rates have been experienced in major European markets and in the US (Lohr 1998).

Europe has the world's largest market for organic food. The value of the organic market in the European Union alone is currently estimated at around $4.5bn (Lohr 1998; PSL 1998). Although this represents less than 2% of the total food market, the figure could rise to $22.5bn (around 10%) by 2006, based on current growth rates (PSL 1998). As with land under organic production, market shares of organic products vary considerably between countries, as do levels of development of the market structure. Table 4.1 summarises market details for selected European countries. The market values given in the Datamonitor report tend to be higher.

With $1.6bn retail sales of organic products, Germany has the largest market for organic food in value terms. This represents 1.5% of total national retail food sales. With shares of over 2% of total food retail sales, however, Austria, Denmark, Sweden and Switzerland have the most developed markets for organic food in Europe. These markets are characterised by a relatively high level of domestic production and have had strong consumer demand since the early 1990s. Consumer awareness is high owing to the major retailers all having effective marketing campaigns for organic food which is available in most multiples. Countries such as the UK, France, Germany and the Netherlands have occupied the middle ground, but in the last two to three years their markets have begun to develop rapidly, partly as a result of the entry of supermarkets and increased availability. In 1996, the annual growth rate of 12% in France was estimated to accelerate to 20% due

Table 4.1 Estimated value of organic markets in Europe, 1997 (source: Datamonitor 1999; GIRA EuroConsulting 1997; Lohr 1998; PSL 1998).

Country	Retail sales $m (estimate)	Retail sales $m†	Retail share % (estimate)	Imports % (estimate)
Austria	270	546	2.5	30
Belgium	75		1	50
Denmark	190		3	25
France	580*	673.4	0.4*	10*
Germany	1600	1789.8	1.5	50
Netherlands	230	281.7	1.5	60
Sweden	200	98.7	<3	30
Switzerland	190		5	no data
United Kingdom	445	419.4	2	70

*1996; †from Datamonitor 1999

to increased supermarket availability (GIRA EuroConsulting 1997). The domestic market for organic food in southern European countries has for many years been negligible with the majority of organically produced products destined for export. In Italy, however, the domestic market is experiencing significantly improved growth rates, although it is still underdeveloped (Zanoli 1998; Datamonitor 1999).

Outside western Europe, there are a number of countries where the market for organic products is becoming more than just a niche market. The US has the largest single-country organic market, valued at around \$4.5bn in 1997, with steady growth over the period 1993–97 (Lohr 1998; Datamonitor 1999). There is also increasing consumer demand in Japan and Australia. Small, but rapidly growing, markets are emerging in less-developed countries such as Argentina, Egypt, Israel, Mexico, South Africa and many eastern European countries, although production in these countries is highly export oriented.

4.2.2 *Key products*

The most important product categories in the European organic market are fresh fruit and vegetables, cereals and dairy, although this varies between countries according to availability of supply, price premiums and consumer preferences. Market shares in each country also vary considerably according to product.

In the UK, fruit and vegetables form the largest share of the organic food market with 54% of organic value sales, followed by cereals with 14% (see Fig. 4.2). However, although demand for organic dairy and meat products has evolved slowly and the values of the markets are much lower, the growth rates for these sectors are much higher than for fruit and vegetables. Between 1992 and 1996 the dairy and meat sectors experienced growth rates of 250% and 188.9% respectively, compared to growth rates of 91.3% for vegetables and 71.4% for fruit (Mintel 1997). The development of demand for these product categories is partly a result of the BSE scare and the increased availability in supermarkets (Mintel 1997; PSL 1998).

The UK situation broadly reflects the situation across Europe although there are of course differences between countries. In a survey of 18 European countries, vegetables were generally considered the most important organic product on the total food market in terms of quantities sold, followed by cereals and dairy products (Michelsen *et al.*

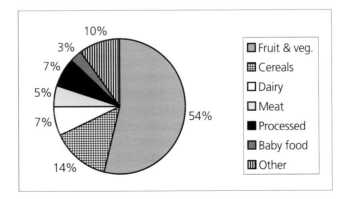

Fig. 4.2 Market share of UK organic retail sales, 1997 Source: Soil Association, 1998.

1999). Table 4.2 summarises some of these product data for selected countries. Only in Austria and the Czech Republic did vegetables not feature largely. Growth rates for the vegetables market vary considerably from 1% in the Netherlands to 58–100% in Greece, Belgium and Switzerland.

Markets for organic animal products (in particular meat) are the least developed, except in Austria and Denmark where organic dairy products form the largest share of the organic market (approx. 9% and 14% respectively). However, as in the UK, markets for organic dairy and meat products seem to be growing at a higher rate than for organic fruit and vegetables and cereals (Datamonitor 1999; Michelsen *et al.* 1999). Significant growth rates for organic beef are being witnessed in France, Sweden and Switzerland. In 1996, the EU organic dairy sector was worth $671.5m and it is estimated that this will increase to $1.75bn by 2002 (PSL 1998). Similarly, the organic beef sector is estimated to increase in value by a significant amount from $311m in 1996 to $1.02bn by 2002.

The underdevelopment of markets for organic animal products may, in part, be due to the lack of an EU standard for animal products similar to the EC Regulation 2092/91, which provides a unifying minimum standard for crop production. Another factor hindering progress in the organic meat sector, in particular for pork and poultry, is the availability of organic feed. In many countries a functioning market for organic feed is practically non-existent, although some countries such as Austria, Sweden and Switzerland have well-established markets (Michelsen *et al.* 1999). As trade in animal products increases, the need for organically grown feed will create a significant market opportunity for arable producers.

In most southern European countries, fruit and vegetables are the most important products followed by cereals, except in Greece where oilseeds (mostly olives) are the most important product on the national organic food market.

Other significant growth areas include processed products, ready meals and snack foods, but the development of these product lines is largely dependent on the availability of raw material. Organic baby food has been particularly successful. In the UK, while organic fruit and vegetable sales represent only 2% of total fresh produce sales, organic baby food accounts for more than 10% of total baby food sales (Soil Association 1998). For Austria and Germany the figures are even higher.

4.3   Distribution channels (market structure)

An organised and efficient marketing structure is a key factor in the development of organic farming. It gives farmers the confidence to convert in the knowledge that they will have access to markets and provides the means by which supply can be effectively distributed to the consumer. The structure of the organic food market has traditionally involved smaller specialised operators. This situation has evolved through a combination of both internal and external factors. The organic sector has long been a niche market which has pioneered small-scale structures within which to operate. Furthermore, the philosophy underpinning organic agriculture has for many involved preserving the links between producers and consumers. Externally, there has been a lack of interest on the part of larger, conventional concerns for whom it has been unprofitable to handle the small quantities involved. However, as demand has increased, these conventional companies see a market opportunity in organic

Table 4.2 Key products in selected countries (source: Michelsen *et al.* 1999; PSL 1998).

	Main products 1996–97	Approx. % share of total domestic food market, 1996–97	Approx. % annual growth rates, 1993–97	Comments
Austria	Dairy Cereals (2)	8–10 2	100–120 100	Meat and dairy: significant amounts sold on the conventional market (approx. 60% and 90%, respectively). Fruit and vegetables: shortfalls exist and price premiums are a barrier to purchase. Other: significant market for baby food and scope to develop the processed food market
Denmark	Dairy Vegetables	14.2 6–10	65–70 (liquid milk only) 30–40	Beef: although organic beef has a very small share of the market, it is growing at a rate of 70%. Fruit: 90% of organic fruit is imported
Germany	Cereals Vegetables	3.4 1.7	10 15	Dairy: 50% of milk is not sold as organic due to marketing problems. Vegetables: over 50% imported. Other: baby food has a relatively high share of the market. Areas with significant potential for growth include fruit and vegetables, dairy, processed foods, ready meals and snack foods
Netherlands	Vegetables Dairy	no data 1	1 0	Demand for vegetables and dairy products is stagnating and there has been a fall in cereal sales. Demand for organic meat is relatively low and despite food scares is growing at a slow rate compared to some other countries. Around 60% of organic fruit and vegetables is exported, but much of this trade consists of re-exports
Sweden	Vegetables Dairy	3–4 2–3	28 77	Growth rates in Sweden are generally high across all product categories, particularly organic fruit (145%) and organic beef (105%). Over 90% of organic fruit is currently imported
Switzerland	Vegetables Cereals	10–12 2.9	58 60	Growth areas include dairy products and, more significantly, beef with growth rates of 65% and 225% respectively
UK	Vegetables Fruit	2.3 1	18 14	70–90% of fresh fruit and veg. is imported. Although they currently occupy marginal shares of the market, growth areas include dairy and beef, with growth rates of approx. 53% and 47%, respectively. Processed foods, baby food and ready meals are also potential growth areas

food and have started to enter the market. This is particularly true of the retail sector where supermarkets are stocking increasing amounts of organic products.

4.3.1 *Processing and wholesaling*

Manufacturers and other operators in the supply chain of organic products (post-production and pre-retail) are SMEs (small and medium sized enterprises) specialising in organic food. They are numerous and highly fragmented, which has resulted in a lack of market transparency in some countries, such as Germany and Italy (Michelsen *et al.* 1999). Among the problems encountered by such companies are the lack of processing capacity and inadequate access to sales channels, and increased price competition. The background to this has been a considerable increase in consumer demand, which has been slow to filter through to the supply side. A situation has developed whereby lack of supply has meant lack of processing capacity, which in turn has led to an underdeveloped marketing structure to distribute the product to the consumer. This has created a lack of confidence in markets among producers, with organic products being sold on conventional markets at conventional prices, which in turn discourages conversion. Undeveloped demand cannot merely be attributed to the poor supply base. Although the German market for organic food is now the largest in Europe, a marketing structure which has not kept pace with increases in both demand and supply has hampered development of the market. During the 1990s, producers have converted in increasing numbers in response to subsidies, creating a situation whereby both demand and supply are increasing significantly, but the increased supply is not reaching the consumer (Hamm & Michelsen 1996). This bottleneck effect, whereby adequate marketing channels do not exist for the increased production, creates excess supply which puts pressure on prices, putting a greater financial burden on an already stretched sector. The excess supply ends up being sold through conventional channels at conventional prices, which in turn inhibits further development of organic marketing channels (Schulze Pals 1994; Thimm 1993).

In addition, a general lack of organisation and co-operation to co-ordinate supply and distribution seems to have prevented the sector from overcoming problems of fragmentation and discontinuity of supply. In some countries, such as Belgium, this has led to increased imports which in turn has meant lower prices and a barrier to domestic conversion (Michelsen *et al.* 1999). In Austria, and in the UK and Danish organic dairy sectors however, greater organisation and co-operation appear to have enabled producers to market more effectively, in particular to the supermarkets which are becoming increasingly significant players. Increased producer co-operation can also give farmers the confidence to convert.

The manufacturing sector is greatly influenced by retailer activity. The major multiples seem to have enabled some specialised wholesaling and processing firms to establish themselves and experience real growth. Many retailers have entered into agreements with processors and wholesalers to try and guarantee supplies. In Austria, for example, Billa has entered into a five-year supply contract with some smaller processors and, in the UK, Sainsbury's supermarket chain has set up the SOURCE group (Sainsbury's Organic Resource Club) with its core organic suppliers and processors to co-ordinate supply.

However, the nature of the relationship and co-operation between the multiples and smaller specialist organic processing and wholesaling firms will inevitably change as competition increases and larger mainstream processors and suppliers enter the organic sector. Several major manufacturing concerns have already responded to the increases in consumer demand witnessed over the last two to three years. Examples include the multinationals Heinz, Unilever and Proctor & Gamble, all of which have one organic line, and more significantly the German baby food manufacturer Hipp, which now uses only organic raw material in its products.

4.3.2 Retailing

A decentralised marketing structure – direct marketing and specialist retail outlets – has predominated and only in the last two to five years have mainstream retailers begun to take a serious interest in stocking organic food.

Supermarkets

Consumer demand has been the driving force behind the entry of conventional companies in the organic marketplace. This is also true of the retail market where supermarkets are becoming an increasingly important sales channel for organic food, particularly in those countries where consumer demand is strong. In Europe, they are the dominant retail outlets for organic food in the Scandinavian countries (Denmark, Finland, Norway and Sweden) and in Austria, France, Luxembourg, Portugal, Switzerland and the UK with a more than 50% share of organic food retailing (Michelsen *et al.* 1999; Datamonitor 1999). Whereas in the UK and France, for example, organic sales are divided between the major multiples, organic retailing in Austria is dominated by one supermarket chain, Billa. Promotional campaigns for its own organic brand have been a feature of Billa's success. Most German supermarkets also have their own labels, but they have only recently made an impact on organic retailing. This is in part due to the dominance of smaller specialist retailers in the German food industry.

Supermarkets operate in a highly competitive market, and so require guaranteed supplies of large quantities of food produced to specific standards at low prices. The inability of the organic sector to meet these requirements led to initial stagnation or slow growth of organic sales in UK supermarkets in the 1980s. Dramatic increases in demand in the late 1980s and early 1990s brought about a change of attitude from UK supermarkets as they began to see a market opportunity opening up. Supermarkets generally have high cosmetic standards for fresh produce which demand a uniform, standardised product as regards size, weight, shape and colour. Recognising that many consumers buy organic produce for 'other' quality components such as nutritional value and associated environmental and health benefits, many supermarkets have introduced different or more flexible quality specifications for organic produce than conventional produce. This ensures greater availability by reducing the amount of outgraded produce (Latacz-Lohmann & Foster 1997). In addition, some supermarkets have sought indirectly or directly to encourage conversion through co-operation with organic producers and suppliers to try and identify their needs and overcome problems in meeting supermarkets' requirements. This can take the form either of informal discussions or, more effectively, through forward contracting over a fixed period. This seems to be a mutually beneficial arrangement that ensures a degree of security for producers, at least in the short to medium term, and secures supply for the supermarkets.

However, it is not only a question of increasing supply. As mentioned in section 4.3, an effective distribution network needs to be in place which can deliver increased supplies to the retail outlets. This can be especially difficult in countries where the organic supply chain is highly fragmented which creates a complex structure, making it very difficult to work efficiently with large amounts.

Specialist shops

In Germany and the Netherlands, small specialist natural food shops which sell organic food and other natural health products are of particular importance, accounting for more than a 30% share of organic food retailing (Michelsen *et al.* 1999). There is no tradition of such shops in the UK, however in recent years some small specialist organic shops have been successfully established. Such specialist shops are independently owned and strongly motivated by organic ideology. For this reason many consumers see them as the only real alternative to buying directly, which can be impractical, especially for those in urban areas. These shops procure their products both direct from the producer and from wholesalers and processors, offering potentially high margins, especially to small-scale producers. Because they are usually small, individually owned enterprises, there is no scope for economies of scale and prices to the consumer tend to be relatively high (Latacz-Lohmann & Foster 1997).

Although they cannot compete with the supermarkets on volume, they offer a wider variety of products. A core group of consumers buy from these shops because of the unique services they offer, such as a less anonymous atmosphere and in some cases greater transparency than the supermarkets. They focus on product presentation and information and it is these service aspects that they need to emphasise if they are to compete with the supermarkets.

Potential competition faced by the multiples exists in the form of specialist greengrocers, bakeries and butchers in Mediterranean countries where there is a strong tradition of shopping in the local shops and markets (Datamonitor 1999). In Greece, Italy and Spain, independent specialist stores account for over 30% of organic food retailing for the majority of the major product categories (Michelsen *et al.* 1999). These outlets are usually well-established and are supplied locally, sometimes on contract, and therefore do not have high marketing costs.

Organic supermarkets

Organic supermarkets are a relatively recent retailing concept in both Germany and the UK. In Germany, their concept varies from store to store, with some creating literally a supermarket which sells organic food where the emphasis is on convenience and practicality, and others creating a whole new way to shop (a hybrid between a supermarket and a the smaller specialist organic shops). Here the customer not only buys everything under one roof as in a supermarket, but also shops in a comfortable environment with access to well-trained, well-informed sales personnel.

Organic supermarkets offer an alternative to specialist organic shops by presenting organic food in an atmosphere with which most people are familiar. It remains to be seen, however, whether and to what extent the organic supermarkets will be able to expand the market as a whole. It has been suggested that only those who are specifically looking for organics will shop at organic supermarkets, and consumers with no particular interest in organics will continue to shop at conventional retail outlets. In this respect, it is doubtful whether organic supermarkets will attract 'new' consumers.

Planet Organic, which opened in 1995, is the UK's first organic supermarket, aiming to attract mainstream consumers to organic food by providing a 'one stop, all-under-one-roof service. Out of This World (OOTW), which opened its third store in April 1996, is the UK's first 'ethical' supermarket chain, stocking a range of products selected for their ethical or environmental value which are locally supplied, fairly traded and/or organic.

Organic supermarkets have a greater chance of economies of scale than smaller retail outlets and, by offering a wide variety in one shop at lower prices to the consumer than smaller specialist outlets, they are able to compete with the supermarkets on convenience.

Direct marketing

The most common forms of direct marketing are farm gate sales, farm shops and farmers' markets and, more recently, box schemes and home delivery. Direct marketing is a well-established marketing strategy for organically produced food, as for many years it was the main marketing option for organic producers.

As with other retail outlets, the importance of direct marketing varies between countries and according to product. Direct sales of fruit and vegetables are important in a number of countries, but of the European Union countries, only Greece seems to directly retail significant amounts (more than 30%) of all the major product categories (fruit 30%, vegetables 40%, cereals 50%, olives 10%, wine 20%) (Michelsen *et al.* 1999). Box schemes are particularly popular in the UK, as are home delivery schemes which are becoming increasingly important in urban centres. Southern European countries such as Greece, Italy and Spain have a strong tradition of direct selling both in the conventional and organic agricultural sectors due in part to the fragmented nature of the retail structure (Datamonitor 1999).

Direct marketing has been an important source of income to organic producers, the alternative being to sell through conventional channels at conventional prices. In 1996, Bioland, the biggest German producer organisation, estimated that approximately 52% of their producers received the majority of their income from marketing their production directly (Latacz-Lohmann & Foster 1997). For the producer (especially the small-scale producer), there is the opportunity for better profits, while at the same time addressing the ethical and environmental concerns of consumers. It also encourages greater links between producers and consumers, and the biggest advantage for the consumer is that they pay less. Moreover, consumers normally obtain very fresh produce and have control over its organic authenticity.

Direct forms of marketing cannot, however, compete with the multiples on convenience. One of the main disadvantages of buying direct appears to be the inconvenience involved, although to a certain extent home delivery services have sought to overcome this limitation. Nevertheless, home delivery and box schemes are generally limited to fresh produce and cannot offer the wide variety or volumes that the multiples and specialist outlets provide. None of the forms of direct marketing will be able to serve the ever-increasing number of organic consumers as organic food becomes more and more mainstream, although this type of retailing will always attract a certain section of organic consumers.

4.4 International trade in organic products

The rate of expansion of a country's organic production does not necessarily reflect the size of a country's organic market in terms of domestic retail sales. Although UK organic production is expanding, for example, demand still outstrips supply meaning that the majority of organic food consumed has to be imported. Similar situations exist in other countries (see Tables 4.1 and 4.3). A number of northern European countries are not self-sufficient with respect to organic production and import more organic products than they export. Key import markets include the UK, where imports represent about 70%, and Germany, where imports account for approximately 50% of the organic market. The Netherlands also has a strong import market, but is unique in that it re-exports a large part of these imports throughout Europe (PSL 1998; Datamonitor 1999). Markets in southern Europe are mostly export oriented.

Imports tend to be highest for fruit and vegetables which cannot be produced domestically. It is estimated that imports account for approximately 90% of total organic fruit sales in Denmark, Sweden and the UK (Michelsen *et al.* 1999). For vegetables the figures are lower, except in the UK, where imports account for 80% of total organic vegetables retailed. Here, large quantities are being imported which could be produced domestically such as potatoes, carrots and brassicas (Soil Association 1998). Imports of organic meat into Europe tend to be much lower, which can possibly be explained by the less developed demand for organic meat among consumers. Another explanation could be the lack up until now of an EU regulation on livestock production which has inhibited trade between countries as products are certified according to different standards, although most international trade adheres to the IFOAM (International Federation of Organic Agriculture Movements) standards for animal production (*Michelsen et al.* 1999).

Organic production in many less developed countries is growing as a result of the export opportunities created by demand, particularly in northern European countries such as the UK and to a lesser extent Germany, where demand is far outstripping supply. The recognition of Argentina in 1993 as a country eligible to export certified organic produce to the European Union, for example, provided a real boost to the development of its production base. Other countries in this category are Australia, Hungary, Israel and Switzerland. In 1997, 74% of Argentine organic production was destined for export, 61% to the EU, 12% to the US (Merlo 1997). Other significant export markets in less developed countries include South and Central America, Mexico, Israel, Egypt, Turkey and some African countries. For the most part, these exports are based on the nature of a consumer

Table 4.3 Imports of main organic products as percentage of total domestic organic food market (source: Michelsen *et al.* 1999).

	Fruit	Vegetables	Cereals	Dairy	Meat
UK	90	70	15	12	<3
Italy	30	no data	no data	80	no data
Germany	56	36	10	6	no data
France	no data	no data	16	20	no data
Sweden	95–100	10–20	1	no data	no data
Denmark	90	25	64	no data	no data

demand that requires a consistent supply of a variety of foods over the whole year, many of which cannot be grown domestically. This is reflected in the product mix of the exports: mostly coffee, bananas, citrus and other fresh fruit, vegetables, sugar cane, cocoa and other products which cannot be produced in many European countries.

Climatic constraints have not only opened up exports markets for countries outside the EU. Spain, Greece, Italy and Portugal all have relatively small domestic markets, but land area under organic production is developing rapidly in response to demand in other countries. The majority of their production (fruit, vegetables, oilseeds – mostly olives – and wine) is exported to northern European countries and beyond (mainly the US and Japan). It is estimated, for example, that 80% of Greek fruit and olive production is destined for export (Michelsen *et al.* 1999). The development of organic production in many of these exporting countries has been supported or encouraged by contracts or projects set up by major retailers and/or traders in the importing country to secure supplies of certain products.

The strong demand for organic food in Europe also presents a significant export opportunity for countries with more established organic production, such as the US, which exports mostly grains, pulses, fruits and a limited range of processed products to northern Europe, Canada and Japan. Canada exports approximately 80% of its production to Europe and the US (mostly grains, pulses and maple syrup; PSL 1998; Beauchemin; 1996). Limited amounts are exported from northern European countries with the exception of the Netherlands which imports and then re-exports relatively large amounts of fruit and vegetables to other European countries.

The export of processed foods is seen as future growth area, especially to northern European countries where demand is growing, but raw materials are still in short supply (Datamonitor 1999). Eastern Europe, particularly EU accession countries, is seen as a potentially significant export market for Europe, especially for frozen berries and seeds (PSL 1998).

4.5 Policy support for organic marketing and processing

As organic agriculture has developed, financial support policies for converting and existing organic producers have been introduced. Such policies have recognised the environmental benefits of organic farming and in part have provided compensation for lower yields and higher costs sometimes encountered during the production process (Lampkin *et al.* 1999). Significant increases in demand have led to a need for a corresponding development of the market structure for organic food in order to keep pace with the growing supply base.

As mentioned earlier with respect to Germany, a niche marketing structure unable to deal with increasing supply can lead to a bottleneck effect whereby supply is not reaching demand. This is a situation to which production-oriented organic aid schemes in isolation from market-oriented can contribute (Hamm 1997). Although production subsidies promote the development of the supply base, in the absence of an effective distribution network they increase competition between producers,and excess supply increases pressure on prices making organic farmers dependent on continued subsidisation. Policies to stimulate production require the existence of a marketing and distribution network able to organise the operation downstream and meet market demand (Lampkin *et al.* 1999).

Support for marketing and processing in the organic sector has been limited. Organic marketing and processing activities in the EU have been eligible for support through EC Regulation 866/90 on improving the processing and marketing conditions for agricultural products and EC Regulation 2328/91 on improving efficiency in agricultural structures. Although one of the priorities for the application of EC Regulation 866/90 is investment relating to organic farming products, few organic operators across the EU have received funding under this regulation and only in Germany has EC Regulation 2328/91 been used, to a limited extent, to fund direct marketing initiatives in Niederschsen (Lampkin *et al.* 1999).

As part of Danish organic aid schemes, financial aid has been introduced not only for production but also for the development of the supply chain and publicity for organic products, as well as information and advice provision for organic producers (Lampkin *et al.* 1999). An evaluation of this development support concluded that projects have been carried out which have resulted in increased trade in organic products. Developmental projects and the dissemination of knowledge and information has resulted in a higher level of expertise in the organic farming sector. This 'professionalisation' of organic farmers and the entry of more full-time farmers has resulted in increased marketing through more mainstream channels as opposed to farm-gate and marketplace sales (PLS 1992).

In Germany and Austria, financial support has been available for regionally implemented programmes. Such regional level support has been of particular benefit to small-scale initiatives and has successfully helped establish and develop regional marketing networks, overcoming the problems of a small organic sector and encouraging the entry of new operators (Lampkin *et al.* 1999). It is doubtful whether such initiatives would have come into existence without public support (Posch 1997). An analysis of support for the sale of regional products through direct marketing in Sachsen concluded that it can increase the number of organic processing enterprises and the volume of processed raw materials. This can have an impact both on the development of the supply chain and conversion as it gives producers and sales and marketing organisations such as distributors the confidence to enter the sector (Jansen 1997).

Some countries such as Germany, have targeted support towards organic producer co-operatives (Lampkin *et al.* 1999). Not only does greater producer co-operation enable producers to offer a greater variety of food and share the costs of marketing investments, it also permits the supply of larger quantities to the multiples. Schmid (1994) suggests that support for producer co-operatives is particularly helpful for new organic farmers in the start-up phase.

A potential barrier to the uptake of some marketing and processing development schemes is restrictive eligibility requirements such as minimum turnover (Lampkin *et al.* 1999). There is also a lack of awareness of the availability of such support and some small-scale operators are deterred by the lengthy and complicated application procedure. In order for support programmes of this nature to be effective, it is necessary to address the specific needs of the organic sector at all levels. Greater provision of information about funding possibilities might also improve uptake of support programmes.

4.6   Future prospects

The organic food market in Europe is forecast to witness buoyant growth, with the sector projected by different sources to account for as much as 10% of total European food sales by the year 2006. Although countries are at differing stages of development and organic sectors differ according to unique country-specific conditions, it is probable that markets will continue to grow steadily. Consumer demand has not reached its limit and supply seems likely to continue increasing in response to that demand.

The markets for fruit and vegetables, cereal and dairy products are the most developed, although there are variations in each country depending on factors such as domestic supply, distribution networks, price premiums and consumer preferences. Key areas for market expansion include the meat sector and the market for processed and convenience foods. The extent to which these sectors will develop depends to a large extent on increased supply.

A major limiting factor to the growth of the market is availability of supply. Some countries in Europe have now set targets for organic agriculture. Austria has already achieved 10% organically managed land area, Sweden and regions of Germany and Switzerland aim to be 10% organic by the year 2000, while Denmark, Finland and Norway have set a lower target of 5%, and France 3%. This anticipated growth in organic production is certain to have significant implications for the development of the markets for organic products (Lampkin *et al.* 1999).

Several of these countries, plus the Netherlands, and Wales in the UK, have developed so-called action plans to provide a framework for achieving such targets. The aim of these programmes is to integrate support for the organic sector by targeting a range of actions which go beyond production to include aspects such as market development, the harmonisation of certification, technical advice and extension, and research. As part of the Agenda 2000 programme, the Rural Development Regulation[1] provides scope for member states to implement similar integrative programmes. The regulation makes specific reference to increasing demand for organic products (Lampkin *et al.* 1999). The tendency towards reinforcing support for organic production with measures devoted to the whole organic supply chain may go some way towards avoiding imbalances between supply and demand and the bottleneck situation mentioned earlier.

At the same time, an effective distribution network is dependent on increases in supply to encourage the entry of new operators and increasing the professionalism of existing operators. Mainstream manufacturers and retailers will become increasingly prevalent in the organic supply chain as they realise the market potential of organic food.

Supermarkets in particular are emerging as the dominant retail outlet in many countries and this trend looks set to continue, subject to the limiting factors of availability and price. Evidence suggests that willingness to pay a price premium in the EU may decline by 2004 (Datamonitor 1999). The exceptions to this are Italy and Spain where the market for organic food is underdeveloped and a higher price premium may be acceptable for longer as the market develops and organic food becomes more popular. However, expansion of the market does not exclude other retail outlets and a diversity of retail outlets for organic food will continue to exist. Increased competition between outlets will require retailers

---

[1] The Rural Development Regulation consolidates existing EU support measures (agri-environment and structural measures) to be implemented through Rural Development Plans designed by each member state. Areas eligible for support include investment aids, agri-environment schemes, processing and marketing of farm products, and the adaptation and development of rural areas.

to be aware of consumers' reasons for buying organic food in their outlet and the type(s) of consumer they are catering for and they will need to promote themselves accordingly (Latacz-Lohmann & Foster 1997).

Based on the principle that a change in quantity results in a change in quality, growth inevitably implies change for the organic movement. Current market growth patterns suggest that there is scope to successfully meet the demands not only of a large, but also a diverse, consumer base. The challenge for the organic sector is to keep in sight the philosophy behind organic agriculture and to communicate this to consumers in order to preserve its distinct identity and to encourage them to continue buying organic food.

References

Beauchemin, R. (1996) Organic Canada. In: *International Organic Market Study* (ed. C. Haest). BioFair, Camar de Comercio de Costa Rica, San Jose.

Datamonitor (1999) *Natural and Organic Food and Drinks, 1999*. Datamonitor plc, London.

Foster, C. & Lampkin, N. (1999) Organic production statistics 1993–1997. Internal report, *Organic Farming and the CAP* project. EU FAIR3-1996-1794.

GIRA EuroConsulting (1997*) Study of the French Retail and Wholesale Market for Organic Food*. GIRA EuroConsulting, Surrey.

Hamm, U. (1997) Staatliche Förderung des ökologischen Landbaus-Absatzfonds statt Flächenpremien. In: *Ökologischer Landbau: Entwicklung, Wirtschaftlichkeit, Marktchancen und Umweltrelevanz. FAL-Tagung 26–27 September 1996* (ed. H. Nieberg), pp. 259–266. Landbauforschung Völkenrode, Sonderheft 175; Braunschweig.

Hamm, U. & Michelsen, J. (1996) Organic Agriculture in a Market Economy. Perspectives from Germany and Denmark. *Fundamentals of Organic Farming, Proceedings 11th IFOAM Conference*, 1. IFOAM Ökozentrum Imsbach, Tholey-Theley, Germany.

Jansen, B. (1997) Erfahrungen zum Einfluß der Förderung auf die Entwicklung des ökologischen Landbaus in Sachsen. In: *Ökologischer Landbau: Entwicklung, Wirtschaftlichkeit, Marktchancen und Umweltrelevanz. FAL-Tagung 26–27 September 1996* (ed. H. Nieberg). Landbauforschung Volkenröde, Sonderheft 175, Bundesforschungsanstalt für Landwirtschaft, Braunschweig.

Lampkin, N & Padel, S. (eds) (1994) *The Economics of Organic Farming: an International Perspective*. CAB International, Wallingford.

Lampkin, N., Foster, C., Padel, S., & Midmore, P. (1999) The policy and regulatory environment for organic farming in Europe. *Organic Farming in Europe: Economics and Policy*, Volume 1, University of Hohenheim, Stuttgart, Germany.

Latacz-Lohmann, U. & Foster, C. (1997) From "niche" to "mainstream" – strategies for marketing organic food in Germany and the UK. *British Food Journal*, **99**, 275–283.

Lohr, L. (1998) Implications of Organic Certification for Market Structure and Trade. In: *Emergence of US Organic Agriculture: Can We Compete?* (eds K. Klonsky & L. Tourte). Principal paper at the American Agricultural Economics Association annual meeting, Salt Lake City, Utah, August 2–5, 1998.

Merlo, S. (1997) Successful Sustainable Development with Government Support. *The future agenda for organic trade. Proceedings of the 5th IFOAM conference for trade in organic products* (ed. T. Maxted-Frost), pp. 12–14. IFOAM and Soil Association, Tholey-Theley and Bristol.

Michelsen, J., Hamm, U., Wynen, E. & Roth, E. (1999) The European Market for Organic Products: Growth and Development. *Organic Farming in Europe: Economics and Policy*, Vol. 7, University of Hohenheim: Stuttgart, Germany.

Mintel (1997) Organic and Ethical Foods, *Market Intelligence*, November 1997. Mintel.

PLS (1992) *Evaluation of Law no 363 10 June 1987 on Organic Farming by PLS-Consult*. Ministry of Agriculture, the Directorate of Agriculture, Copenhagen.

Posch, A. (1997) Making growth in organic trade a priority. *The future agenda for organic trade. Proceedings of the 5th IFOAM conference for trade in organic products* (T. Maxted-Frost), IFOAM and Soil Association, Tholey-Theley and Bristol, pp 9–12.

PSL (1998) *The European Organic Food Market*. Final Report for the United States Department of Agriculture. Produce Studies Ltd, The Hague.

Schmid, O. (1994) Agricultural Policy and Impacts of National and Regional Government Assistance for Conversion to Organic Farming in Switzerland. In: *The Economics of Organic Farming* (eds N.H. Lampkin & S. Padel), pp. 393–408. CAB International, Wallingford.

Schulze Pals, L. (1994) *Ökonomische Analyse der Umstellung auf ökologischen Landbau*, Schriftenreihe des Bundesministeriums für Ernährung, Landwirtschaft und Forsten, Vol. 436. Landwirtschaftsverlag, Münster.

Soil Association (1998) *The Organic Food and Farming Report, 1998*. Soil Association, Bristol.

Thimm, C. (1993) Neue Absatzwege und Elastizität des Marktes. *Ökologie & Landbau* **86**, 39–43.

Zanoli, R. (1998) Ökologischer Landbau in Italien. In: *Ökologischer Landbau in Europa*. (ed. H. Willer), pp. 198–216. Deukalion, Holm.

5 Fruit and Vegetables

Robert Duxbury

5.1   Introduction

To many consumers, the opportunity to buy fresh organic fruit and vegetables is foremost in their minds when they make the choice to eat organic foods. Fresh organic produce represents many of the fundamental issues which draw consumers to make the decision to eat organic, whether it be a concern about pesticide residues, a desire to support sustainable farming systems or a deeper understanding of what organic production actually represents. There are many reasons why people make the organic choice and the decisions that they make when buying fresh produce takes them right to the heart of what they perceive, or actually know, about organic foods.

Broadly speaking, there are two fundamental reasons why people buy organic foods. The first is an internalisation of the reasoning; that is to say, people are interested or concerned about the food that they eat. They want to ensure they make a choice which gives them the best option for health, safety, nutrition, taste and flavour. The old saying 'you are what you eat' is most apt to this group of people. They could also be called 'foodies'. The foodies may represent an important growth area in the organic market, particularly in terms of the relationship between organic foods and taste and flavour – many people insist that, subjectively, organic food 'tastes better'.

The second reason for people making that organic choice is related to external issues. They are interested about how their food is produced, where it comes from and what is done to it in the process of getting it from the field to their plate. They will be concerned about environmental issues, the sustainability of agriculture, the countryside and the future of the world. This group of people could be called 'greenies'. The greenies represent many of the intellectual choices being made for buying organic food, having interests in environmental issues and concerns about conventional farming.

Of course these definitions are oversimplified and the reality is that many consumers will be a mixture of foodies and greenies, often relating their choices to their own personal experiences or needs. We are all continually making choices which transcend both these internal and external forces. It is interesting to note that the history of organic production and the standards is very much based upon, and determined by, the producers, many of whom were involved right from the early days. However, one can begin to detect a greater degree of interest and influence coming from the consumer end, as the organic market becomes more sophisticated. Indeed it is the pressure of consumer choice that has so dramatically projected the organic world forward over the last five years.

We know that much of the recent compunction for consumers to choose organic food has been driven by media attention to 'food scares', which have popularised the focus on food safety and health. Issues such as salmonella in eggs, *E. coli* food poisoning and the BSE disaster have all taken their toll on consumers' confidence in the food industry.

At the same time these issues have underscored and publicised the choice of organic foods as a viable alternative.

But one further and, arguably, the most significant factor that has focused consumers minds on organic foods over the last two years has been the issue of genetically modified organisms (GMOs). Again, media attention has highlighted the choice offered by organic foods, as the only viable alternative for those consumers who do not wish to buy food derived from GMOs.

This chapter seeks to explore the background related to organic fruit and vegetables and throughout it will try to keep a perspective which relates directly back to the consumer, not forgetting the foodies and greenies. The following pages look at some of the major issues confronting growers, producers and certifiers in the expansion of the organic fresh produce sector. It will assess the current market situation and deal with a selection of individual crop or production situations which are causing a degree of debate and discussion within the organic arena. It will also tackle some of the concerns being raised over perceived limitations, framed within organic standards, in terms of how far producers and retailers can go in widening the range and availability of organic produce.

Throughout this chapter there will be an obvious view taken from the supermarket's perspective. Not only is this chapter written by a supermarket representative, but multiple retail sales of organic foods account for 69% of the total UK sales of organic foods, estimated to be worth £269 million (Soil Association 1999a).

5.2   The market

Sainsbury's has seen a 30-fold increase in organic sales since 1997 when it had just 47 organic foods available. Sales are currently approaching £3m weekly with over 500 organic lines. With the total organic market estimated to be worth around £500m by the end of 1999, Sainsbury's now represents about 30% of all organic food sold in the UK. The range, seasonality, variability and versatility of fresh produce is unsurpassed by any other area of organic food. Sainsbury's alone sells over 300 different fruits, vegetables and salads, of which some 60 lines are organic. This produce is sourced from over 50 countries around the world.

The total market for organic produce in the UK is currently worth about £175m, of which some 82% is imported, according to the Soil Association's *Food and Farming* report (1999a). The report places the fruit and vegetable sector as having the greatest share, followed by multi-ingredient foods, dairy and cereals. The wholesale market value of home produced organic fruit and vegetables is estimated to be £34m (see Table 5.1), although the report indicates that overall UK demand (40%) is growing faster than supply (25%).

Table 5.1   UK Market Value of Organic Food (source: Soil Association 1999a).

Destination/Use	Market value £m.	Percentage value
Direct, farm-gate sales and box schemes	58	15
Independent and multiple retailers	332	85
Total	390	100

As far as UK production is concerned, the Soil Association's *Food and Farming* report estimates there to be about 3000 hectares of organic fruit and vegetable production in the UK (see Table 5.2) which represents only 5% of the total production area. This includes protected cropping.

A particular issue, more directly concerning fresh produce rather than some of the other areas of organic food supply, is that of imports. More than 70% of organic foods sold in the UK are imported (Soil Association 1999a). When it comes to fresh fruit and vegetables, that figure averages out at 80% and, for particular items, such as tomatoes, may be closer to 95%. In the case of fruit, many products simply cannot be grown in the UK, or, for that matter, even in Europe. Bananas are a good example. There is a clear demand for organic bananas and this fruit is the single most popular organic produce line sold by Sainsbury's. Sales doubled in 1998 to £1.5m and organic bananas now represent 2% of total banana sales.

The principal desire of any multiple retailer is to offer their customers range, availability and value. The driving force is to justify why the customer should shop at a particular supermarket rather than choosing that of a competitor. That means that fresh produce has to be available throughout the hours of opening, be attractively presented and of best quality and freshness. The ultimate desire for some retailers is to have an organic alternative for each conventional product and, in some cases, actually replace the line with a wholly organic one. Waitrose, for example, is pursuing this policy with a number of lines now entirely organically grown.

The issue of whether organic food should be imported is a question that is hotly debated. The term 'food miles' identifies the concern that some people have about the distance food has to travel before it gets onto the plate. In the modern global market, with its sophisticated distribution systems, fresh produce can be conveyed virtually anywhere within 24 hours. However, the true cost of such activity cannot readily be calculated. If an airliner is flying regularly between, for example, Nairobi and London, why should it not convey freshly grown legumes or fruits in its hold? If the African farmer can earn more money by growing crops for export, who will judge that as wrong?

Much has been written about the global economy and the concept of sustainability. Organic standards seek to encourage production systems that work at a local level, drawing upon local resources and producing food for local people. They also seek to minimise pollution and to reduce energy use, particularly of non-renewable fossil fuels. Clearly a conflict exists and that is being exacerbated by the complete lack of demand-met production in the UK at the present time. The whole world knows that the UK market wants more organic food and the whole world is ready, willing and able to supply it.

Table 5.2   UK organic fruit and vegetable production (source: Soil Association 1999a).

Crop	Area under cultivation (ha)	Estimated tonnage	Value (£m)
Green vegetables	991	7300	4.41
Potatoes	911	17500	5.25
Roots	495	8350	2.79
Protected crops and salads	208	11000	5.6
Mushrooms	no data	1600	3.7
Fruit	395	2951	2.07
Total	3000	48701	23.82

Another area of concern, linked to that of imports, is the issue of seasonality. There is a strong feeling on the part of some consumers that fruit and vegetables should retain seasonal identity, and that the aim by supermarkets to have every produce line represented 365 days of the year is misplaced. The supermarket's response to this is that choice is offered to customers who then have the discretion of opting to buy or not. The question is, however: does this conflict with the basic ideals of organic production?

Storage is another subject which raises concern. The concept that all fruit and vegetables are 'fresh' has to be understood in the context that storage plays an important part in all aspects of agriculture. The tradition of bringing in the harvest and storing food is as old as agriculture itself. If a farmer has only one opportunity to grow a single crop each year, it is vitally important that he can store it for use throughout the rest of the year. It must be remembered that processing is just another method of storage. The processors will often use frozen fruits and vegetables as their principal raw materials.

The issue of appearance as it relates to the cosmetic perfection demanded by consumers and supermarkets alike is a fundamental point. It is obvious that even organic customers will buy on appearance. Misshapen, blemished and deteriorating fruit and vegetables represent less value for money and more preparation time. It must also be remembered that the Grading Standards for Fresh Fruit and Vegetables mandate the need for a minimum Class II attainment for most (but not all) types of fresh produce. However, many consumers of organic fruit and vegetables are equally interested in taste and flavour.

Supermarkets are clearly not the only source from which consumers choose to buy their organic produce. Indeed, the organic world presents unique opportunities for consumers to buy produce that does not exist in the conventional arena. Box schemes and farmers' markets represent a different and interesting way of buying fresh fruit and vegetables. In addition, farm shops are becoming an increasingly popular outlet for produce.

5.3   Availability

(See Table 5.3.) In the world of fresh fruit and vegetables, there are three cardinal requirements: price, quality and availability. Availability brings together the requirements of price and quality in terms of the requirement for continuity, which in turn can be demonstrated as consistency of supply. Both retailers and processors need some guarantee of availability if they are not to disappoint their customers. Clearly, the purchase of any crop is subject to an agreed specification and growers will have to bear the requirements of that specification in mind when planning their crops.

However, there is a further dynamic at work here which is much more critical in the organic than in the conventional sector. It is the need for adequate programming that is critical to the success of supplying the organic market. This requirement must start with the farmer or grower. The imperative here is that farmers and growers are becoming more mindful of the need to grow a crop to a specific market requirement, not only to reduce financial risk but also to ensure the supply chain has a balance of availability. This factor is further compounded by the requirements that organic farmers and growers have to fulfil balanced rotations on their land. The complications of carefully worked out rotations which can span a 5-year or even 7-year period mean that long-term marketing plans must be set up to deliver availability.

Table 5.3 Principal organic fruit and vegetables: availability by country of origin and time of year (source: Sainsbury's supermarkets).

Crop	Country of origin	Jan	Feb	Mar	Apr	May	Jun	Jul	Aug	Sep	Oct	Nov	Dec
Root vegetables													
Beetroot	UK	*	*	*					*	*	*	*	*
	Holland	*	*	*	*			*	*	*	*	*	*
	Spain			*	*	*	*	*					
Carrots	UK	*	*						*	*	*	*	*
	Spain				*	*	*	*					
	Italy			*	*	*	*	*					
	Israel		*	*	*								
	Holland	*	*	*					*	*	*	*	*
Garlic	France								*	*	*	*	*
	Spain	*	*					*	*	*	*	*	*
	Egypt					*	*						
	Argentina	*	*	*	*								
Leeks	UK	*	*	*	*					*	*	*	*
	Holland	*	*	*	*			*	*	*	*	*	*
Onions	UK	*	*						*	*	*	*	*
	Spain	*						*	*	*	*	*	*
	Italy	*	*						*	*	*	*	*
	Argentina			*	*	*	*						
	Holland	*	*	*	*				*	*	*	*	*
	Egypt					*	*	*	*				
Parsnips	UK	*	*	*						*	*	*	*
	Spain				*	*							
	Holland	*	*									*	*
Potatoes	UK	*	*	*	*		*	*	*	*	*	*	*
	Spain					*	*	*			*	*	*
	Italy					*	*	*					
	Israel	*	*										*
	Holland	*	*	*	*				*	*	*	*	*
	Egypt	*	*	*	*								
Swedes	UK	*	*	*	*				*	*	*	*	*
Brassica vegetables													
Broccoli	UK							*	*	*	*	*	
	Holland							*	*	*	*	*	
	Spain	*	*	*	*	*						*	*
Cabbage – green	UK	*	*	*	*	*	*	*	*	*	*	*	*
	Holland	*	*	*	*	*	*	*	*	*	*	*	*
Cabbage – red	UK	*								*	*	*	*
	Holland	*	*	*	*	*			*	*	*	*	*
Cabbage – white	UK	*	*	*	*				*	*	*	*	*
	Holland	*	*	*	*	*			*	*	*	*	*
Cauliflower	UK	*	*	*	*			*	*	*	*	*	*
	France	*	*	*	*	*					*	*	*
	Spain	*	*	*	*							*	*
	Italy	*	*	*	*	*				*	*	*	*
	Holland	*	*	*	*				*	*	*	*	*
Sprouts	UK	*	*							*	*	*	*
	Holland	*	*	*						*	*	*	*

Table 5.3   (*Continued.*)

Crop	Country of origin	Jan	Feb	Mar	Apr	May	Jun	Jul	Aug	Sep	Oct	Nov	Dec
Other vegetables													
Asparagus	UK						*	*					
	Spain			*	*	*							
	Chile									*	*	*	*
	South Africa	*	*	*									
Aubergines	Holland				*	*	*	*	*	*	*	*	
	Spain	*	*	*							*	*	*
	Italy	*	*	*	*							*	*
Beans and peas	UK						*	*	*	*			
	Holland						*	*	*	*			
	Denmark						*	*	*				
	Egypt	*	*	*	*	*	*				*	*	*
	South Africa		*	*	*	*	*						
Courgettes and/or	UK							*	*	*			
marrows	Holland							*	*	*	*		
	Spain	*	*	*	*	*						*	*
	Italy	*	*	*	*						*	*	*
	Egypt	*	*	*	*	*					*	*	*
Fennel	Spain	*	*	*					*	*	*	*	*
	Italy	*	*	*	*				*	*	*	*	*
Squash	UK							*	*	*	*	*	
	South Africa	*	*	*									
Sweetcorn	UK								*	*	*		
	Spain			*	*	*	*						
	France							*	*				
	Holland								*	*			
	South Africa	*	*	*	*								
	US		*	*	*	*							
Fresh herbs	UK	*	*	*	*	*	*	*	*	*	*	*	*
	Spain	*	*	*	*	*	*	*	*	*	*	*	*
	France	*	*	*	*	*	*	*	*	*	*	*	*
	Holland	*	*	*	*	*	*	*	*	*	*	*	*
	South Africa	*	*	*	*	*	*	*	*	*	*	*	*
	Israel	*	*	*	*	*	*	*	*	*	*	*	*
Mushrooms	UK	*	*	*	*	*	*	*	*	*	*	*	*
	Holland	*	*	*	*	*	*	*	*	*	*	*	*
Protected salads													
Capsicums –	Holland				*	*	*	*	*	*	*	*	
green/red/yellow	Spain					*	*	*	*	*	*	*	
	Italy	*	*	*						*	*	*	*
	Egypt	*	*	*	*	*						*	*
	Israel	*	*	*	*	*						*	*
Cucumbers	UK						*	*	*	*	*		
	Holland					*	*	*	*	*	*	*	
	Spain	*	*	*	*	*					*	*	*
	Italy	*	*	*	*							*	*
	Egypt	*	*	*	*	*						*	*
	Morocco	*	*	*	*	*						*	*
	Israel	*	*	*	*	*						*	*

Table 5.3   (*Continued.*)

Crop	Country of origin	Jan	Feb	Mar	Apr	May	Jun	Jul	Aug	Sep	Oct	Nov	Dec
Tomatoes –	UK					*	*	*	*	*	*	*	
(inc. cherry and	Holland				*	*	*	*	*	*	*	*	
on the vine)	Spain	*	*	*	*							*	*
	Italy	*	*	*									*
	Morocco	*	*	*	*	*						*	*
	Israel	*	*	*	*	*	*	*				*	*
Root salads													
Radish	UK					*	*	*	*	*	*		
	Holland	*	*	*	*	*	*	*	*	*	*	*	*
Spring onions	UK					*	*	*	*	*			
	Egypt	*	*	*	*	*						*	*
	Mexico	*	*	*	*	*							*
Leafy salads													
Celery	UK					*	*	*	*	*	*		
	Holland				*	*	*	*	*	*	*		
	Spain	*	*	*	*	*	*					*	*
	Italy	*	*	*	*	*						*	*
	Israel	*	*	*	*	*							*
Chinese leaf	UK									*	*	*	
	Spain	*	*	*	*	*						*	*
	Italy	*	*	*	*							*	*
Iceberg lettuce	UK					*	*	*	*	*	*		
	Holland					*	*	*	*	*	*	*	
	Spain	*	*	*	*	*						*	*
	Italy	*	*	*								*	*
Little gem and	UK						*	*	*	*	*		
cos lettuce	Holland						*	*	*	*	*		
	Spain	*	*	*	*							*	*
	Italy	*	*	*								*	*
Protected lettuce	UK					*	*	*	*	*	*		
and other outdoor	Holland	*				*	*	*	*	*	*	*	*
	Spain	*	*	*	*							*	*
	Italy	*	*	*	*							*	*
	California	*	*	*	*	*	*	*	*	*	*	*	*
Watercress	UK					*	*	*	*	*	*		
	Portugal	*	*	*	*							*	*
Deciduous fruit													
Apples	UK									*	*	*	*
	Argentina		*	*	*	*	*	*					
	Chile		*	*	*	*	*						
	France	*	*	*	*				*	*	*	*	*
	Italy	*	*	*	*				*	*	*	*	*
	Germany	*	*	*	*				*	*	*	*	*
	New Zealand		*	*	*	*	*						
	US	*	*	*	*						*	*	*
Pears	UK									*	*	*	*
	Argentina		*	*	*	*	*	*					
	France	*							*	*	*	*	*
	Italy	*	*						*	*	*	*	*

Table 5.3   (*Continued.*)

Crop	Country of origin	Jan	Feb	Mar	Apr	May	Jun	Jul	Aug	Sep	Oct	Nov	Dec
Stone fruit													
Peaches and	France							*	*	*	*		
nectarines	Italy							*	*	*	*		
Citrus fruit													
Lemons	Italy	*	*	*	*	*	*					*	*
	Spain	*	*	*	*	*	*					*	*
	Greece	*	*	*	*	*	*					*	*
	Turkey	*	*	*	*	*						*	*
	Uruguay						*	*	*	*	*		
Oranges	Italy	*	*	*	*	*	*					*	*
	Spain	*	*	*	*	*	*					*	*
	Greece	*	*	*	*	*	*					*	*
	Morocco	*	*	*	*	*						*	*
	Uruguay						*	*	*	*	*		
	Argentina						*	*	*	*	*		
Easy peelers	Italy	*	*	*	*	*						*	*
	Spain	*	*	*	*	*						*	*
	Morocco	*	*	*	*	*						*	*
	Uruguay						*	*	*	*	*		
	Argentina						*	*	*	*	*		
Soft fruit													
Strawberries	UK						*	*	*				
	Spain					*	*	*	*	*			
	Holland					*	*	*	*	*			
	Argentina	*	*	*									*
Bananas	Dominican Republic	*	*	*	*	*	*	*	*	*	*	*	*

More and more supermarket buyers are working to programmes which, in turn, can give the farmer or grower a clear idea of anticipated demand. This is important to the producer as he has to plan his rotation sometimes two or three years ahead, deciding on which crop will follow on in a particular field. He will have to order seed (an issue which holds further complications in the future for organic producers) and ensure that his fertility building and nutrient budgeting meet the requirements of the target crop within the rotation sequence. In horticultural and, especially, within stockless rotation systems this becomes an appreciably demanding exercise.

Processors too are having to respond to the market by giving their producers more precise programmes. As the market becomes more sophisticated it will lead to processors demanding specifications which may not necessarily be met simply by grading a range of product from a crop primarily grown for the fresh market. Increasingly crops are being grown for a specific use and, indeed, the performance of particular varieties will have a direct effect on the end use requirement. It is the issue of varietal performance and availability of seed which is set to become a major challenge to organic growers as the requirement for seeds themselves to be grown organically becomes part of organic standards.

The opportunity to import out-of-season fresh fruit and vegetables gives further opportunities to widen availability as we have seen in section 5.2. The programming of imported sources must match with the continuity of home supply. The requirement for European importers to comply with increasingly complicated import regulations, particularly from non-EU countries, is discussed elsewhere in this book.

As the harmonisation of standards worldwide progresses, it is hoped that we will see the development of systems that can demonstrate equivalency become more generally accepted. IFOAM's International Organic Accreditation Service could play an increasingly important role in this area (IFOAM 1999).

5.4   Growing systems

The methods of production of conventional fruit and vegetables have changed dramatically over the last 50 years. Concern over the intensive nature of many of these production systems has led directly to the development of organic production. However, conventional production continues to change and the most profound development over the last five years has been the introduction of integrated crop management (ICM), with most UK growers now signed up to a production assurance scheme of this type. This development has been led by the UK supermarkets in response to their customers and will have a significant impact on the way in which most conventional crops are grown, not only in the UK but also worldwide.

The importance of this should not be underestimated in terms of organic production. It can be argued that it was the pressure of the organic lobby, which brought about the impetus that led to the introduction of ICM. While ICM is clearly a long way from organic production, a view might be taken that a kind of reverse pressure is now beginning to be seen. As conventional production 'cleans up its act', the challenge for organic standards is to maintain an obvious differential between organic and conventional production systems.

An example of where this challenge will be keenly met is in the requirement that organic crops must be grown from organic seed. At first sight this requirement appears overly restrictive and will certainly cause problems in terms of volume availability and varietal diversity. However, when looked at in the overall context of the development of the organic world, it can be the only way forward in the maintenance of the differential between organic and conventional production systems.

The fundamental requirements for growing organic fruit and vegetables are contained within the organic production standards. Those standards require that production, generally, is soil based and that growing techniques will seek to limit the need and quantity of external inputs, listing those inputs approved for use with organic production. Organic production must substantially rely on farm or locally derived renewable resources. The background for this is framed in Regulation 2092/91 and the details are contained in Annex I and Annex II of that Regulation (MAFF 1999).

Annex I outlines the principles of organic production at farm level and these principles are further amplified through the United Kingdom Register of Organic Food Standards (UKROFS). The standards of individual certification bodies such as the Soil Association contain further requirements. Annex II is split into two sections. Section A deals with materials and products for use with soil conditioning and fertilisation while Section B lists those products approved for plant protection (Soil Association Organic Marketing Co. 1999).

5.4.1 *Fertility*

Annex IIA of the Regulation places a great importance on the use of animal manures and correct composting practices (MAFF 1999). The source of material for composting and animal manures must themselves be derived from organic systems wherever possible. The important point here is that the demands upon the growing of fruits and vegetables mean that such crops will need fertility enhancement in order to produce crops of suitable yield and quality.

The principal requirement for any growing crop is the balanced and accessible availability of nitrogen. However the standards do accept the possibility of so-called 'stockless' systems, where crops may be grown without the use of animal manures while deriving their nitrogen requirements from other sources.

One of two principal ways in which this may be done is by growing a green manure crop as a fertility-building break within the rotation. Normally a legume species is grown which is then incorporated into the soil to provide nutrients and to enhance soil fertility and structure for the following crop. The other way in which fertility and soil structure may be built up, without the use of animal manures, is by using certain natural fertilisers which are also listed in Annex IIA. These substances include products such as rock phosphate and calcium sulphate. However, many of these may only be used under restricted conditions where a specific need can be demonstrated, perhaps where a certain mineral or trace element deficiency in the soil has been recognised.

There has been some criticism of organic systems due to concerns about the risks of pathogenic contamination of fruit and vegetables from animal manures. The Soil Association has drawn up guidelines which stipulate correct techniques for composting which ensure adequate times and temperatures to destroy pathogens as well as weed seedlings and other potential residues (Soil Association 1999b). There are also restrictions on the timing of applications of manures to land on which food crops are grown and this ensures that there can be no contamination of the produce when it reaches the point of harvest.

5.4.2 *Pest and disease control*

Annex IIB of the Regulation lists the products and substances permitted for use in controlling pests and diseases in organic crops. Again the standards make a point of the fact that organic production will encourage healthy plants. If the crop has been grown correctly, it will have developed a natural resistance to pest and disease attack. But the philosophy goes further behind this by looking at a holistic approach to production. Correct rotation, the avoidance of mono-cropping, the provision of a wide spectrum of biodiversity and the build-up of natural predators all work together to avoid crops reaching a point of imbalance and susceptibility.

It has been a point of great discussion that there is the 'treadmill' of the nitrogen cycle into which conventional growers are forced, which determines the pressure of using increasingly more plant protection products. By encouraging a faster crop growth rate with the use of artificial nitrogen, plants become inherently 'soft' and consequently more prone to pest and disease attack. The application of a pesticide often results in a check to crop growth, which may be compensated for by further application of nitrogen. This initiates a see-saw effect in which plants become stressed and vulnerable.

Organic systems seek to have a more measured pace of crop growth which encourages a more robust plant to resist pests and diseases. The greater balance of biodiversity achieves a more natural level of pest predators and a non-stressed crop is more likely to resist disease. Unless specifically targeted, some pesticides will kill beneficial as well as pest insects.

Organic standards are sometimes criticised for allowing certain substances in Annex IIB. These substances are deemed to be 'natural' chemicals not derived from the development of modern agrochemicals such as organophosphates or organochlorines. It must be noted that there is a rigorous approvals system that any plant protection product has go through before it can be legally used. In the UK the Pesticide Safety Division (PSD) of MAFF is responsible for this.

Pesticides approved for organic production also have to comply with the PSD requirements. However, it must be stressed that the use of any plant protection product on an organic crop is seen very much as being a matter of last resort. Growers need to seek derogations for their use from their certification body. The EU is currently undergoing a complete review of all pesticide approvals and there is a strong likelihood that substances such as copper and sulphur will be dropped from the approvals list in future.

Some concern has been raised over the lack of harmonisation of the approved pesticides in organic standards between different states. There are some plant protection products available to organic growers in Holland and France which cannot be used in the UK. Clearly this is disadvantageous to British growers and should be resolved at the earliest opportunity by the Commission.

5.4.3 *Weed control*

There are no approved chemical methods to control weeds in organic crops. Growers have to rely primarily on preventive or physical measures to maintain weed control. These measures include rotation, timing of cultivation, undersowing and the use of mulches. Physical methods may also be used and these include hand weeding, the use of mechanical techniques such as brush hoes and thermal destruction.

The impact of weed control in organic production must not be underestimated. In vegetable production particularly, weed control will often pose more of a challenge than pest or disease control. It is the cost of controlling weeds that can have a significant effect on the economics of organic vegetable production.

5.4.4 *Conversion*

The minimum time period for conversion to fruit and vegetable production will be two years. For open-field vegetable producers this period will be relatively straightforward, as the land in conversion will undergo a fertility building phase. This conversion phase may be also seen as a buffer period, in the case of land which was previously being intensively cropped, giving time for any previously accumulated chemical residues to degrade and dissipate.

The government has provided for funds to support growers converting to organic farming under a grant aided initiative called the Organic Farming Scheme (Lampkin & Measures 1999). However, there are some cases in which the conversion period may cause particular hardship to growers. This may be particularly difficult for top fruit growers who are converting an established orchard. They will not be able to claim the organic premium for their fruit during the conversion period, but at the same time will possibly suffer yield and quality reductions as they discontinue the use of artificial fertilisers and crop protection products.

Another sector of the fresh produce industry that may be disadvantaged during the conversion period is glasshouse production. Some glasshouse producers will not even qualify for grant aid due to the comparatively small area of land being converted. Put together with the high capital costs incurred in setting up a glasshouse unit, there could be a discouragement here for such growers to embark upon conversion.

One possibility is the option to sell 'in-conversion' produce. This option is formally recognised in the organic standards, but it does open the door to criticism of a two-tier or halfway house market. Certainly for top fruit growers, the financial support that they could derive from having a slight premium for in-conversion fruit, could make the difference as to the viability of undertaking conversion. This also gives consumers an opportunity to seek out and purchase in-conversion produce, thereby directly supporting those converting growers.

5.4.5 *Site selection*

Clearly, the demands of fruit and vegetable production are not always suited to every organic farm. Vegetable production generally requires a reasonably level site which is not too exposed, with a soil which is not too heavy but well drained. For brassica crops, the pH level can be a major factor, although liming may be possible to increase pH levels. The accessibility of the site will be important, particularly for cultivations and during the winter months if vegetables need to be harvested in all conditions.

The history of the site will determine how to best undertake the conversion period as well as the following rotation, particularly in terms of weed control. Fruit growers will have to consider the site in terms of geography and climate. Generally, the viability of UK top fruit production is marginal north of a line between the Bristol Channel and the Wash. Frost risk, local humidity and temperature profiles will need to be taken into account. The availability of water for irrigation will also be a factor and many growers are now installing reservoirs and irrigation systems.

For protected crops, the siting of glasshouse structures will need to be assessed in terms of ambient light levels as well as considering the greater demands upon soil type posed by protected cropping. The additional concern over the pest and disease history of the soil will need to be carefully assessed because soil sterilisation will not necessarily be an option for glasshouse producers as steaming is likely to be removed from the standards in the future.

5.4.6 *Seed availability and plant propagation*

The requirement for all organic crops to be grown from organic seed will present considerable challenges to the organic vegetable grower in particular. Currently there is a derogated allowance that conventional seed may be used, but that derogation ceases on 31 December 2003. It might be the case that the derogation to use conventional seed will have to continue with certain species.

Currently only natural seed may be used, with any sort of seed dressing being prohibited. Some growers are concerned about the loss of quality and vigour in seed, particularly the potential increase in seed-borne diseases and viruses. This could quickly cause problems over a few generations of production and create further problems with the health of organic crops.

All propagation of organic plants for transplanting out must take place under strict organic conditions. This means that the compost in which they are sown or rooted has to be an approved organic medium. In addition, there must be adequate separation between conventional and organic propagation where this activity takes place at a conventional propagating enterprise. The availability of plant protection products for propagation is the subject of current review. As we have already seen, the approvals system for fungicides in particular needs to recognise the specific requirements for organic plant raising.

5.4.7 *Harvesting and storage*

One of the most critical aspects in the production of fruit and vegetables is harvesting the produce. The potential to spoil the crop and end up getting a lower return is very acute. It is important that growers consider how they will harvest a crop before they grow it. Many crops require specialised techniques to ensure they reach market in the right condition. The most obvious factors are adequate field heat removal, chilled storage and distribution. Shelf life and performance will be severely curtailed if the produce is not subjected to the correct temperature controls.

A crop which illustrates an example of such demands upon post-harvest conditions is bulb of onions. The requirement here is that the crop is subjected to a sequence of drying and curing in order to preserve the bulbs and to retain an acceptable skin finish. This becomes more important with organic crops because growers cannot use storage chemicals, like maleic hydrazide, which prevents onions from bolting prematurely. Potatoes also need to be handled carefully because the storage fungicides and sprout inhibiting chemicals available to conventional growers cannot be used on organic crops.

Storage facilities need to be kept clean and sound. If a store has been used for conventional crops previously, it may be that there are residues from storage chemicals remaining in the structure or fabric. Bulk bins or boxes should be dedicated solely for use with organic crops, as, once again, the residues of storage chemicals can be very persistent.

5.4.8 *Packing and presentation*

Any operation involved in the grading, packing and distribution of organic fruit and vegetables will need to be certified and subject to regular inspections. This is also true for enterprises involved in processing or the preparation of fresh produce for further processing. There is a strong view that such operations should be solely dedicated to organic production and that there is always a risk of cross-contamination associated with dual-activity enterprises.

In the UK the Soil Association takes the view that pack houses handling organic fruits and vegetables should be solely dedicated to organic production. They will allow a limited transitional period, during which time a new entrant to the organic market may undertake dual activity within the same building. Clear labelling and product traceability are key requirements and there are some very good quality assurance systems now available which make the task of batch identification much more effective and efficient.

As far as supermarkets are concerned, the presentation of organic fruit and vegetables is generally in a pre-packed format. This is partly to do with the need to retain the integrity of the organic produce as well as give information to customers on the labelling. It is not necessary for supermarkets themselves to register as an organic operation if they stick to simply displaying organic produce. However customers do raise the issue of why their organic produce has to be packed or overwrapped and we could see the development of loose or free-flow organic fruit and vegetables in supermarkets in the future.

5.5   Conclusion

The most critical point to be remembered when reading this chapter (indeed throughout this entire book) is that the integrity of organics is everything. It is not enough to allow for regulation, inspection and certification to support organic integrity. It is beholden on everyone involved in this unique area to do their utmost to maintain absolute integrity.

Without that integrity, organic foods will be for nothing in the eyes of the consumer; the concept of 'clear blue water' between organic and conventional production must be maintained. Indeed it is clear that the need for the continuous improvement of the organic standards will be an ever-present requirement.

As integrated farming systems, the reduction of pesticides and integrated pest management become standard practice in conventional agriculture, the challenge for the organic world will be to maintain that 'clear blue water'. The inevitable result will be an ever tightening and further definition of organic standards. Without this, consumers will loose confidence in that integrity and fail to understand what lies behind their reasons for choosing organic foods They might not recognise why organic foods do more accurately represent the true cost of production.

References

IFOAM (1999) *Basic Standards for Organic Agriculture and Food Processing*. International Federation of Organic Agricultural Movements, Tholey-Theley, Germany.

Lampkin, N. & Measures, M. (1999) *Organic Farm Management Handbook*. University of Wales, Aberystwyth and Elm Farm Research Centre, Newbury.

MAFF (The Ministry of Agriculture, Fisheries and Food) (1999) *MAFF Consolidated Version of Council Regulation (EEC) No 2092/91*. Her Majesty's Stationery Office, London.

The Soil Association (1999a) *The Organic Food and Farming Report*. Soil Association, Bristol.

The Soil Association (1999b) *Guidelines for Manure Use Within Organic Systems*. Soil Association, Bristol.

The Soil Association Organic Marketing Company (1999) *Standards for Organic Food and Farming*. Soil Association Organic Marketing Company, Bristol.

6 Organic Meat and Fish: Production, Processing and Marketing

Richard Maunder with Bob Kennard

6.1 Organic meat

6.1.1 *Introduction*

Until the recent explosion of interest, organic food was generally to be found and associated with health food shops and vegetarianism. Organic meat sales were slower to gather pace because they did not fit easily with this type of outlet. However, over the 1990s the profile of the average organic shopper has changed, so that organic meat is now to be found more commonly among the wide range of organic products on offer at the supermarket or at the specialist organic retailer.

Organic meat has only really seen a substantial increase in demand since the mid-1990s, and fuelled by continued food scares the mainstream consumer has been looking for more assurance that food is healthy and safe to eat. In 1987 the retail sales value in the UK of organically produced food was around £40m. Today that figure stands in excess of £390m, of which meat products have a farm-gate value of £7m (Soil Association 1999).

Interestingly, the new organic shopper does not buy exclusively organic food, but a mixed basket. It would appear that some young families buy organic products for their children, but then eat conventional food for themselves. The reasons for this may be poor availability of organic mainstream products, personal preference, or price, but it is important to learn the reasons if we are to assess the longevity of the current surge in organic food sales.

Today's consumer is more sophisticated and has greater choice than ever before. Purchasers are not only seeking meat that is healthy and wholesome, but are increasingly concerned about the welfare of farmed animals. This demand has been largely met by free-range standards of production, but organic farming takes standards and principles further, to a more inclusive and all-embracing level. However, although the consumer is showing greater interest in all facets of meat production, the ultimate price of a product on the shelf or in the farm shop will play an important role in the final purchase decision for the majority.

While demand for organic meat currently exceeds our ability to supply from the UK, land in conversion to organic status now exceeds the amount already converted. By April of 1999 there were over 240 000 hectares of organic and in-conversion land in the UK, which accounts for just 1.3% of the total agricultural area. Of this area, 60 000 hectares had gained full organic status and was certified to produce organic food, while the other 180 000 hectares is within the conversion period (Soil Association 1999). The shortfall in supply is therefore being made up with imports, and if demand continues at the current rate, imports will continue to be necessary in the medium to long term.

Soil is the constant factor in all farming operations, and livestock play an important role in the natural balance of the mixed organic farm. Ideally livestock and crops form a rotational alliance on the organic farm, to return to the soil the manure and plant residues that will be recycled and used again. One of the fundamental issues in deciding on the level of involvement in organic livestock is the ability to provide sufficient organic feed. While beef and lamb may be predominantly grass fed, organic pork production requires large quantities of prepared feed. There are huge demands on organic cereals from other sectors, and this has driven feed prices to unprecedented levels (£260 per tonne for pig rations and £295 per tonne for poultry rations, in October 1999).

This chapter has been contributed by a collection of people who are directly involved in the practical application of bringing organic meat to the marketplace. It will take the reader broadly through the issues from organic production standards, livestock production and marketing, processing standards and requirements, marketing outlets, and finally a brief review of the state of organic meat production in the rest of Europe.

6.1.2 *Production standards*

The word 'organic' is now a legal definition when applied to food and drink products. To qualify, all products have to be certified by a government-approved body, registered with the United Kingdom Register of Organic Food Standards (UKROFS). The Soil Association (SA) is increasingly becoming the dominant and best known of the certifying bodies, however there are others such as Organic Farmers and Growers (OF&G), the Biodynamic Agricultural Association (BDAA), Irish Organic Farmers and Growers Association (IOFGA) and the Scottish Organic Producers Association (SOPA).

Livestock standards relative to beef, sheep and pigs do vary marginally between the various certifying bodies, and therefore the broad principles are outlined rather than providing a detailed survey of the various certifying body standards. A key objective of organic agriculture is to sustain livestock in good health by adopting effective management practices, including high standards of animal welfare, appropriate diets and good stockmanship, so that remedial treatments become increasingly unnecessary.

Origin of animals

Livestock systems should be planned so that stock are born and raised on an organic unit, and the breeds of livestock should be selected to be suitable for raising under local conditions. These animals can then be used in the meat chain.

Animals may be brought in from a non-registered source within a 10% annual allowance, where for example there is a desire to change breed or expand, providing care is taken to source healthy stock with full traceability and a full historic record of medical treatment. However, this stock is subject to the following conversion criteria:

(1) Suckler cows must undergo a conversion period of not less than 12 weeks immediately prior to calving, during which time they must be managed organically.
(2) Ewes must be put to the ram on the organic holding and farmed organically.
(3) Sows must be mated on the organic holding and farmed organically.

Animal health

An important objective of organic livestock husbandry is the avoidance of reliance upon the routine or prophylactic use of conventional veterinary medicines. Instead, good animal health and the prevention of disease is promoted on the organic farm by the adoption of high standards of animal welfare, appropriate housing and feeding systems, lower stocking densities, a high standard of stockmanship and the use of a wide range of alternative treatments and medicines. Rotational clean grazing is fundamental to maintaining good animal health.

Organic farmers are still expected to seek conventional veterinary advice and treatment when serious animal health problems arise. Animal welfare is paramount and veterinary treatment must never be withheld where there is a serious health risk. UKROFS permit the use of anthelmintics (wormers) where individual animals are showing signs of carrying an unacceptable worm burden, and the use of vaccines where there is a known farm problem with specific diseases which cannot be controlled by any other means.

Veterinary medicines must be used in accordance with their UK licence and as directed by a veterinary surgeon. Withdrawal periods between the end of the treatment and marketing of the animal for meat, must be not less than 14 days and/or three times the recommended time, whichever is the greater. Any meat from animals treated with growth promoters, artificial hormones, or organophosphorus and organochlorine pesticides cannot be sold as organic.

Following on from the BSE crisis, UKROFS included standards that related to this subject. These standards state: 'In herds where animals have contracted BSE (bovine spongiform encephalopathy), or where animals have been brought in from herds in which BSE has occurred within the previous six years, then all contemporaries and first-generation progeny of all BSE cases must be removed from the herd and must not be sold as organic.' Contemporaries are defined as animals originating from the same herd, which shared the same food, or were born into the same herd, or were born within three months either side of the date of birth of the BSE case.

At the time of writing there has not been a single reported case of BSE in an organic herd, and this must be regarded a successful outcome of the organic system.

Welfare and housing

Conditions on the organic farm must conform with the highest possible welfare standards. Housing and management must be appropriate to the behavioural needs of the animal, for example, the availability of straw for resting at parturition (birth of offspring) in sows. This in turn should ensure healthy and contented livestock, which should lead to productive animals.

The MAFF codes for animal welfare apply as the absolute minimum standard, but it is useful here to quote some of the UKROFS standards:

'Stalls and cubicles in which animals are confined individually only while feeding are permitted providing the animals have free access to them. No animal should normally be housed out of sight or sound of others of its own species. Animals must not be subjected to any surgical or chemical interference which is not designed to improve the animal's own health or well-being or that of the group. Castration and de-horning are permitted where it is judged to be necessary and in accordance with the relevant Codes of Recommendation for Animal Welfare.'

Livestock diets

In an ideal organic world all organic feed would be produced on the farm and would be sufficient for all the livestock, thus providing perfect balance and harmony. However, the reality is that few farms are self-sufficient, and in fact 'organic farm UK' is far from self-sufficient and much of our organic feed is imported from around the world. Such is the shortage of organic feed that existing rules allow a percentage of non-organic to be included, purchased from approved sources (see Table 6.1). This has led to much debate as to whether this is compromising the organic ideal to satisfy commercial means.

The UKROFS standards state that an absolute minimum of 50% of the dry matter intake must be organic, but that the balance up to the minimum organic part of the diet may be sourced from registered 'in conversion' holdings (where the producer is farming organically but has yet to be licensed by a certifying body), for example, a beef diet could be 50% organic, 40% in-conversion and 10% non-organic.

Organic beef and sheep should be fed according to the traditional methods of grazing when suitable, and hay or silage during winter or wet months. All feed must be GMO (genetically modified organisms)-free, and any livestock which had received any feed which contained GMO ingredients would be ineligible for sale as organic.

Nutritious soil is important to the organic farm, as is the grass management system employed. Sward management is key, and grass at the right length for maximum nutritional benefit is fundamental to the success of the organic system. Some people use a sward stick to help gauge the optimum time for stock to go in or out of pastures. This then assists with worm control and clover growth.

Livestock records

There are general MAFF codes of record keeping, but organic requirements are enhanced in two areas:

(1) Where livestock is bought in, the source and number of animals, any previous quarantine measures and any conversion periods should be recorded.
(2) All feedstuffs including the amounts of non-organic material and the sources must be recorded.

Table 6.1 UKROFS non-organic feed allowance for meat-producing livestock, calculated on a dry matter basis.

Livestock	% Non-organic feed allowance
Beef	10
Sheep	10
Pigs	30

6.1.3 *Livestock production*

Livestock are an important ingredient for the well-being and maintenance of the organic mixed farm. The fact that there has been less interest in livestock than in produce has probably been due more to the size of organic enterprises and issues of feed requirements, rather than to the importance of livestock to the balanced organic farm. Many of the early organic enterprises were relatively small, and produce offered a more viable means of earning a living from the farm.

This situation is changing rapidly as more farms convert to organic principles, and livestock production is becoming more prevalent. Whether organic producers farm beef, sheep and/or pigs will depend on the type of farm, position and climate. The upland farm will have a very different production plan to the lowland farm, but with sensible and logical co-ordination both types of farm can help bring more finished livestock to the marketplace.

It is known that each season many organic store sheep are sold as conventional due to an inability to tie in with lowland holdings with organic pastures readily available. The dilemma is that the organic trade is still very young, and the marketplace for trading stores needs to develop. There are also concerns over the potential likelihood of importing disease from another farm. At present there is no easy method to establish a 'going store price', and the deals are negotiated individually between producers. There is effectively no visible marketplace for these animals. However, the organic marketplace is growing rapidly, and these structural issues will resolve themselves in due course when participants increasingly recognise the opportunities.

Traditionally, many arable farmers would have had a livestock finishing enterprise. The livestock would have provided the fertility for the farm, and the hill farmer would have had an outlet for store animals. Intensive farming after World War II, with increased inputs and attractive subsidies, discouraged the traditional livestock finisher. Farming crops on a large scale was altogether less hard work than raising livestock. Crops such as turnips would have been used to build soil fertility and to have finished store lamb ready for marketing. In the meantime hedges have been removed to create larger arable fields, thus creating an environment alien to livestock, and the whole ecosystem of the farm has deteriorated.

The typical family-run mixed farm in the south-west of England is ideal for conversion to the organic principle, and thus it is no surprise that there have been many conversion applications from this part of the UK. Data from the Organic Conversion Information Service (OCIS) show that 25% of enquiries have been registered from the south-west. One finds that often the grandfather and grandson of the family generations are taking an active interest in organic livestock farming, while the intervening generation is part of the postwar thrust to increase yields and maximise the use of available subsidy payments.

Beef production

Organic beef production is predominately based on the suckler cow system. The breed type will depend on the nature of the farm and its system. Commonly the breeds must be good grass finishers, such as Limousin, Hereford, Angus or South Devon, and not the classic intensive finishers such as the Belgian Blue or the Charolais, although crosses of these terminal sires with native females can be successful in producing meat suitable for the modern organic consumer.

The suckler cow is chosen for its good milk production and mothering ability, while the 'calves at foot' gain resistance to the worm burden and infection by disease. The management of the cow is a key component of successful rearing, and the fitter the animal the easier the calving. The typical 12-month calving index would see the 9-month gestation period followed by 3 months of rest. Weaning at 8–10 months would allow 2–4 months of rest prior to calving again.

Clean grazing is critical for worm control, and grass quality from good management systems will help finish the beef over a 18–23-month period. Silage cut early in the year provides nutritious fodder in the winter months when stock are often brought inside. Usually stock would be brought inside in November and then let outside again in April, depending on the conditions prevailing. Yarded cattle require buildings with good ventilation for cover, reducing the risk of illness such as pneumonia.

Organic suckler cows could have a lifespan of 15 years or more. Should there be a calf fatality at birth, a non-organic calf may be sourced in to take off the cow's milk, but this calf cannot be sold into the organic meat chain.

Sheep production

Sheep are an integral part of the livestock unit as they are such a useful management tool. Organic sheep farming is not much different to conventional systems, but sheep have extra value to the organic system by virtue of their versatility and grazing method. They keep the farm tidy, keep docks under control and help build grass sward strength, while fertilising naturally as they graze. Healthy grass then helps keep weeds at bay, and by virtue of sheep not breaking the soil up in wetter conditions, weeds are not able to take hold.

Given that finished lamb for slaughter is the main output product, much thought must be put to the type of breed that best suits the farm and its climate, with a view to the best marketing opportunities. While the marketplace is looking for lamb all year round, this demand will need to be met from a variety of farms with their own unique systems and seasonal marketing patterns.

Typically the flock would be rotated around clean grazing among the typical 5-year rotational farm plan, and normally the flock would not revisit the same grassland for the consecutive year. The clean grazing policy should minimise the build-up of internal parasites.

Up to 10% of replacement stock can be bought in and taken in to the organic system, with the remainder of replacements sourced from the annual ewe lamb crop.

Pig production

Organic pork production has until fairly recently been a small-scale enterprise on the organic farm, due to the difficulties of formulating sufficient rations at a viable price. As the demand was relatively small for both pork and poultry rations, feed compounders have been sitting on the sidelines watching the market develop.

Fortunately some progressive companies have recognised the potential strong demand and have been endeavouring to meet the demand for organic rations. The difficulty they have faced is a lack of interest from the arable sector to convert to organic status, and competition for sources of raw materials from around the globe. However, organic pork production is now showing the greatest rate of increase in the fresh meat sector, with consumer demand increasing, and once land is converted pork production can be relatively rapid compared to lamb and beef production.

Traditionally the low numbers of organic pigs on farms have been the more traditional breeds, such as the Gloucester and Berkshire, which have good mothering ability and have produced good fresh pork weights albeit perhaps on the fatty side. An expanding organic market is looking for a slightly heavier pig which could also supply the bacon and ham market without putting too much fat down. Research is under way (at Aberdeen University) to learn more about the optimum breeds and crosses which best suit organic production. Although the Large White crossed with the Duroc might fit some farming systems and land type, it would be unwise to generalise when farming enterprises can differ greatly.

Broadly, the pattern is to wean at 8 weeks of age, following a 16-week gestation, and to allow a week for the sow to get in pig again, adding up to a 25-week cycle. The organic sow could average 18–20 piglets reared per year. The pigs would fatten outside for most of the year, but would come in to strawed yards during the winter months, depending on soil type and climatic conditions. Typically pigs would be marketed at 22–26 weeks, according to customer requirements.

The organic system employed to grow pigs will vary from farm to farm, from the classic 5-year rotational system with pigs moving on to clean grazing each year, to the farm that is focused primarily on produce and the pigs follow in behind the crop. Produce and pigs are complementary, as pigs tidy up any produce that is unable to be lifted, or misshapen or under-sized produce that fails to meet customer specifications.

Continuity of feed supply is a major issue for the pig farmer, and it is hoped that more arable farmers will be encouraged by the current demand to grow organic crops on a sustainable rotation. As organic feed is in short supply, it is likely that organic pork will be significantly more expensive than conventional pork for some time.

6.1.4 *Livestock marketing and transportation*

Organic principles ideally suggest that livestock should be marketed to the nearest slaughtering outlet, and be transported directly from the farm to the lairage (livestock receiving point), prior to slaughter, in the belief that the shorter journey leads to reduced stress in transit. But slaughtering centres are in continued decline, and livestock increasingly have to travel further to the nearest slaughter point. An accredited organic centre could be even further away. However, the most stressful part of a journey is the loading and unloading, and therefore the length of the journey (within reason and depending on the species) is perhaps not as critical as the quality of livestock handling during the whole process. The use of trained livestock hauliers and good-quality bedding on the vehicle ensures that high welfare standards during transit can be maintained.

The slaughtering industry has changed dramatically during the 1990s, and it is now some years since there was a slaughterhouse in every town. Increasing legislation and economic rationalisation has led to a dramatic decline in the number of slaughterhouses, however, this has increased the potential ways to market organic livestock. Many more slaughtering centres are now accredited for organic slaughter and thus able to apply the symbol of the certifying body.

Direct marketing to a slaughter centre

Many producers prefer to have a direct relationship with the slaughterhouse, and to receive payment direct from them, normally within 10 days of kill. The benefits of this type of relationship are numerous.

(1) Within a short time of the livestock being processed, the producer is able to obtain the grade results. This can be extremely beneficial. For instance, if results show the stock to be either too heavy or too fat, the producer is able to draw stock again quickly (particularly with lamb).
(2) The producer is encouraged to see stock being killed and to discuss the grades and quality of the stock with the slaughterhouse.
(3) Discussion enables producers to understand the requirements of the slaughterhouse in forward planning, to translate these into forward opportunities for marketing, and to maximise these benefits.
(4) The producer has the opportunity to get closer to the ultimate customer and their requirements.

Indirect sale through a marketing co-operative

Some producers prefer to be part of a marketing co-operative. There is a cost of belonging to such a co-operative but this has to be set against the potential benefits:

- Being part of a wider marketing planning organisation;
- The feeling of being part of a bigger organic family, with the confidence and networking it might bring;
- Distribution of surplus store animals should be a key function of a producer group, adding real benefits to the producer.

A direct relationship with a local butcher or farm shop

This form of livestock marketing might occur where a local butcher selects from the field, offers a price, liveweight or deadweight, and arranges the slaughter for the producer. This type of relationship can keep costs to the absolute minimum, and can be a very useful outlet for the producer.

Electronic marketing

This form of livestock marketing has not been used to any great degree for organic livestock, mainly due to the relative lack of organic livestock available to market. It involves offering up livestock to a company that is running an electronic auction with a series of buyers who are all online, competing against each other. It could be of interest to organic producers because the livestock are lifted from the farm and do not go via an auction mart. However no relationship is really formed with the buyer, and thus this method might be regarded as too remote for the organic farmer. In the future this medium could be useful for marketing store stock to a wider audience.

Auction markets

This form of marketing is rarely used for organic stock, apart from store stock to be sold as conventional. This is partly due to the lack of stock, but also because these channels do not sit easily with organic ethics, and are usually ruled out by certifying bodies. Welfare may be compromised by additional and unnecessary transport and handling, and other organic standards can more easily be compromised in the environment of a predominantly conventional livestock market.

6.1.5 *Processing: standards and application*

All establishments which process organic products and plan to utilise an organic symbol, whether slaughterhouse, meat processing plant, or smaller retail butcher, must have a certificate of registration that is issued by an authorised recognised body, for example the Soil Association, or the Organic Farmers and Growers Association.

It is very important if both conventional meat and organic meat are produced on the same premises to protect the organic integrity of that product. Organic processing runs are separated from conventional runs by either space or time, that is either by using a dedicated processing line or by ensuring that organic processing only occurs at defined times. Cleandown procedures must fulfil organic requirements. Working procedures must be recorded for verification and annual audit.

Where a plant operator is producing beef that is labelled 'organic', under the Beef Labelling Scheme which came into force on 1 July 1998 and became compulsory on 1 January 2000 (in the UK but not the rest of Europe!), the operator is responsible for traceability and for the validity of all labelling claims, and will be audited by a third party to verify historical records and management systems to ensure they are robust enough to validate the claims made.

Traceability of the organic product is extremely important; processors must have a system in place which enables traceability from the retail pack purchased by the customer back to the farm batch. This applies to any slaughterhouse, butcher or retail packer, whatever the size of the organisation. Traceability is a core component of organic processing, and is stringently audited to ensure consumer confidence in the final product.

Slaughter

On delivery to the slaughterhouse the documentation that arrives with the animals must be verified, and the organic licensee checked to prove the validity of the licence number. Once data checks are completed, the farmer's name, address and licence number are transferred onto a working document which starts the traceability system. The paperwork must then be forwarded to the lairage. At Lloyd Maunder, for example, this traceability sheet is referred to as the 'passport system'.

On delivery to the lairage, livestock are inspected by the on-site veterinary surgeon or the Meat Hygiene Service. This is known as the ante-mortem check, and the animals are examined for signs of disease, injury, fatigue or stress, and checked for cleanliness. All animals should be kept in the lairage within their social groups. All farm stock are slaughtered within farm batches, and these batches are given a reference number which is then recorded on the passport. If the animals are being slaughtered in a conventional meat abattoir then the organic animals are the first animals of the day to be slaughtered; if for any reason this is not possible then all machinery and equipment must be washed down before organic production can be started.

After the slaughter, dressing and evisceration process, the carcasses are then inspected by the Meat Hygiene Service, to ensure the meat and offal is fit for human consumption in accordance with the Fresh Meat – Hygiene and Inspection – Regulations 1995 (as amended).

After inspection the carcasses are weighed and graded and the organic symbol is applied by an independent assessor, usually the Meat Hygiene Service or on-site veterinary surgeon, who holds the organic stamp. Any carcasses not stamped with the organic symbol should not be used for organic meat production. Carcasses are then placed in chillers designated for organic carcasses. If it is impossible to use a designated chiller then the organic carcasses must be clearly separated from other products. A slow chill regime for the first 10 hours is sensible to help prevent cold shortening of the meat. It would be usual to hold lamb for at least 48 hours to mature, and beef for at least a week hung on the bone.

Once through the required chill regime, the passport information must be checked and completed and sent with the carcasses to its next destination. Temperatures of carcasses leaving the slaughterhouse should be monitored and recorded; temperatures must be no higher than 7°C . All loading of carcasses must take place within covered loading bays. When delivering organic carcasses with conventional meat, then the carcasses must be kept separate at all times. All paperwork, passports, and any other information taken during process runs must be kept for a minimum of three years.

Cutting

On receipt of the organic carcasses, all the documentation which arrives with the product must be checked. In particular the carcasses must have been stamped with their organic symbol, and have a reference number attached for traceability to continue through the primal cutting process.

The temperature of the cutting room should be no more than 12°C, and once the carcasses have entered the boning room they should be butchered as quickly as possible, to provide optimum vacuum-packed life. Once all the carcasses with the same reference number have been primal cut, they then need to be vacuum-packed and labelled. This needs to be completed before the carcasses with the next reference number are brought into the cutting room to be primal cut, to reduce the chance of batch mixing.

The description label to be placed on the vacuum-pack bag must state that the meat is organic. It must include the reference number which relates to the place of origin and also the date of packing and maximum life date within which the product must be retail packed or sold. Beef is often given up to 4 weeks' window life in the vacuum bag to improve maturity and tenderness, as long as it is held at suitable temperatures (0°C).

The passport system gives everyone from the abattoir to the retail packer and on to the retail outlets the information required to produce a quality product with full traceability from the farm to the consumer purchase point, which also meets the Soil Association and government requirements.

6.1.6 *Further processing of meat*

As the supply of fresh organic meat improves there is increasing opportunity to develop more innovative products, particularly as the range of ingredients that can be supplied organically is now growing rapidly.

It has long been recognised that it is vitally important to utilise every part of the carcass, and this is particularly true with organic meat where premiums paid for the meat must be recouped from a full range of interesting products. The importance of developing a range of products that can utilise trim is vital when developing a range for marketing that might otherwise consist of just the prime cuts.

Pork has always been an extremely versatile meat with products such as sausages, hams and bacons complementing the fresh meat offer. Beef and lamb likewise can utilise trims into organic minces, burgers or grill steaks as well as various cuts in the sliced cooked meat arena. Given the premium paid for organic meat, it is important to be innovative with product ideas, and not assume that organic status is sufficient to guarantee consumer loyalty. Products must be exciting and relevant to the modern lifestyle.

The manufacturing sector is an area of marketing which is currently seeking large quantities of organic meat-based ingredients. Organic chilled and frozen ready meals are beginning to reach the market, and soon the convenience market will provide a wider organic choice. Ingredient and sauce manufacturers have been historically reluctant to produce organic products, but this is changing with the scale of consumer demand and producers' ability to produce in volume.

6.1.7 *Marketing of organic meat and meat products*

To market any product one has to know the customer. Who is the organic meat customer? How many are there? What do they want? How often do they purchase? What is their average spend? Where do they want to buy from? How far will they travel? Do they buy fresh, frozen or fresh for freezing? What products sell to what age profile? Do they always buy organic? Do they buy some organic products or mostly organic?

Unfortunately far less market research has been carried out for organic products than for conventional. This makes the task of marketing organic meat a great challenge. To date, marketing has been largely production led; the consumer has had to look for the product. Until recently there has been no major brand informing the consumer of the merits of supporting the purchase of organic meat. However the main supermarket brands are now taking up this role, and it will not be long before major international brands become more established in the marketplace.

Despite this lack of focus the organic market continues to expand, driven by the lack of growth and capital returns from the conventional market and a hard core of producers seeking fundamental change to the way that food is produced in the UK.

The wide choice of marketing opportunities open to a producer, agent, processor or retailer are greater than ever before. The producer has a variety of options as to how to market their product in a way that is of most benefit to them. They can stay as a producer, or extend their involvement in becoming a vertically integrated business, controlling all parts of the supply chain in marketing the product from the farm direct to the consumer. Farm-gate selling has been, and will continue to be, a part of the organic meat offer. It has the advantage of allowing the producer to be in control of their product from plough to plate. The traceability of the product and the 'feel good' factor to the consumer of visiting the farm are its key marketing strengths. Producers are dedicated to organic food in its total concept and in marketing terms are selling the complete article.

Supermarkets in the UK now account for 75% of all meat purchases. To be mainstream, organic products have to be offered in supermarkets as an alternative to the conventional product across all product ranges. Small producers with a fragmented supply base are difficult to market effectively to supermarkets. The policy of supermarkets is to work with a few dedicated suppliers in partnership to provide their customers with good-value, quality product all year round. The lack of continuity of supply is the reason that some supermarkets entered the market and came out again in the early 1990s, and those who stayed in with a long-term view have found the market difficult to grow. However the current interest in organic food is now paying dividends for those supermarkets who remained loyal to the sector as they have been well placed to capitalise on their connections.

There are more marketing approaches to the customer than ever before. Because the market is new, it is well set to embrace new routes such as the internet, farmers' markets and buying co-operatives. The arrival of the internet has opened up an enormous opportunity for organic farmers to market their products directly to customers, and it has enabled consumers struggling to find organic meat to locate a suitable supplier and buy direct. Search engines such as Yahoo.com can allow potential customers who key in words such as 'organic' and 'meat' to access any number of e-commerce sites with which to do business.

6.1.8 *European perspective*

The trend towards healthy eating across Europe is now well established and customers are embracing the benefits of organic food. Demand for organic meat is outstripping supply in most European countries. The most developed markets are in Austria, Germany, Sweden and Denmark. Even within these markets there is considerable scope for further growth and penetration. The key to success depends on organic being able to compete with the conventional meat market on price expectations, taste and range.

The Austrian organic food market is more advanced than the UK because of the domination of the market by multiples who account for 80% of all organic food sales, compared with 35% in Holland and as low as 10% in Spain. In Austria, organic food purchase now accounts for nearly 10% of all food purchases. Multiples play a key part in organising and developing the market for organic meat (European Natural and Organic Food and Drink 1999).

Holland, by comparison, has been supplied with organic fresh meat by a network of specialist organic butchers all purchasing from a central abattoir to provide continuity of supply and standards. Because of the butchers' strength in this part of the market, some of the large supermarkets are starting only now to offer organic fresh meat.

Currently one of the major challenges facing organic meat producers is to ensure that organic standards exist on an international basis. IFOAM (International Federation of Organic Agricultural Movements) plays an important role in determining that in the future there may be a common baseline of standards around the world. There is growing concern within the organic network over this issue. With the pressure on supply to meet the UK demand, imports are flooding in to the country under various labels. There is a fear that the introduction of other certification systems might lower the price of organic meat and undermine the home product which is produced to a higher standard.

In countries such as Sweden, where there is one organic certifying body (KRAV), there is a clearer focus on producing a common standard definition of organic. This makes the marketing of organic far easier in labelling and building customer sales and confidence. In the UK we have a variety of 'organic marks' which can be confusing to the customer.

The future fresh meat offer across Europe will maintain high growth levels and see further development of organic meat into the convenience and ready meal sector. Children's food is a large potential growth sector. So far the organic meat offer has largely been in simple cuts. The lack of innovative new products in the organic meat sector could open the gap for multinational brands to enter the market with their substantial marketing budgets. These are indeed fast-moving and exciting times for those involved in the organic sector.

6.2   Organic fish

6.2.1 *Introduction*

Fish have only recently made an appearance on the organic scene, but with concerns over some aspects of the production of conventional fish, it is likely that organic fish will have a significant impact on the market.

Organic fish are divided into two groups – farmed and wild. The European organic standards only recognise farmed fish, and once these standards are introduced, wild fish will be relegated to the description of Soil Association Certified.

6.2.2 *Farmed organic fish*

At present only a handful of fish farms are registered as organic in the UK. They produce either salmon or rainbow trout. The conventional fish farming industry has become highly intensive, requiring the use of a number of chemicals. However, with this type of farming has come cheap salmon – a prospect inconceivable only a few years ago.

In order to become organic, fish farmers have to ensure that no pesticides, colourants or other chemicals are used in the production of their fish, and that the fish have plenty of room to move about. Most fish farming entails the use of cages, in estuaries or lakes or at sea. The same applies to organic production, but stocking densities must be considerably lower. Finally, the feed given to organic fish must fulfil the same criteria as feed given to any organic livestock, including no drugs or other additives. The pink coloration found in salmon and trout must originate from natural sources.

Farmed salmon and trout are available for sale in various forms – whole gutted fish, fillets and smoked. Salmon is available in the familiar cold-smoked form, whereas trout is available in the hot-smoked form.

6.2.3 *Wild organic fish*

There is only one source of certified organic wild fish, and that is the remote island of St Helena in the South Atlantic. This source was identified by Graig Farm Organics from mid-Wales, who arranged for certification of the fish by the Soil Association, and who now import the fish to the UK.

To be certified, the fishing system must fulfil a number of criteria :

- They must be fished from unpolluted waters (St Helena is one of the most remote places on earth, with no industry);
- There must be a minimal environmental impact (all St Helena's fish are caught by hook and line);
- There must be proven sustainability (the St Helena fishing fleet is about 20 strong, all family-owned small boats);
- There must be complete traceability of the fish (each fisherman's catch is recorded by the St Helena Fisheries Corporation);
- There must be distinct social benefits to the local community (all processing is carried out on the island, ensuring that as much added value as possible is retained on the island).

The St Helena range consists of steaks of yellowfin and albacore tuna and wahoo (a member of the Barracuda family), fillets of mackerel (a different species to the North Atlantic variety – smaller and less oily), grouper (with a white, flaky, cod-like appearance), bullseye (a delicate, fine-textured fish), lobster (or crawfish) tails, together with a range of smoked fish (hot- and cold-smoked yellowfin tuna, and cold-smoked wahoo).

Acknowledgements

Special thanks are due to many who contributed to this chapter: producers David Ursell, Graham Vallis and Cyrill Blackmore; staff at Lloyd Maunder: John Bailey, Clive Griffiths and Carolyn Batten.

References

Blake, F. (1981) *The Handbook of Organic Husbandry*. Crowood Press, Ramsbury
EEC (June 1991) EEC Council Regulation No. 2092/91.
Soil Association (1999) *The Organic Food and Farming Report 1999*. Soil Association, Bristol.
UKROFS (September 1997) Standards for organic food production.
Wright, S. (1994) *Handbook of Organic Food Processing and Production*. Blackie Academic, London.

7 Organic Poultry Meat Production

Peter Challands and David Lanning

7.1 Introduction

Forty years ago chicken was considered a luxury for Sunday lunch and now it is a commodity always available on the supermarket shelf. Over these years we have witnessed the intensification of poultry rearing, and with this has come criticism of the industry, some of it justified and much that is probably unjustified. Whatever else is levelled at intensification, one cannot deny that poultry meat has provided a high level of nutrition to many families at prices that can be afforded by most. Standards of husbandry and welfare have moved forward and present-day poultry farmers are very aware of their responsibilities in producing a quality product.

In the UK, chicken is the number one meat for the tenth consecutive year and in 1998 accounted for 38.4% of the primary meat market, its nearest competitor being beef at 24.2%. The total retail chicken market was valued at £1.65bn in 1998, an increase of 2.8% over 1997 (British Chicken Information Service 1999).

Chicken is no longer a product that is simply produced in the UK for domestic consumption. It is now imported from many countries of the world. It is a truly internationally traded commodity. This world trade is not the only change which the poultry meat market has experienced: there has been a shift from frozen to fresh, a change from whole chicken to portions, the never-ending demand for breast meat and away from dark meat. Finally, there is the growth in the value-added sector, which in 1998 increased its share of the total retail chicken market from 36.3% to 38.6% of all sales (British Chicken Information Service 1999). Where does organic chicken meat feature in this changing scenario?

Currently there is a far greater demand for organic poultry meat than there is supply, which reflects its increasing popularity with consumers. This trend is creating a whole new market for poultry meat. It appears that some consumers require a wide range of high-quality chicken from suppliers using sustainable, high welfare, and environmentally sound farming practices. Does it signal a trend away from intensification and into a different style of farming, or will it achieve a particular share of the market and eventually slow down? Does it reflect a need for improved taste, a new 'in mouth entertainment?' Perhaps the 40 years of rapid development of the poultry industry requires yet further refinement as part of the evolution of the chicken market.

The many food scares which occurred in the 1980s and 1990s, as well as increasing reports of allergies and dietary disorders, may have hastened this change in attitude. Consumers may need to feel certain that the food they eat is safe and that organic is the way to arrive at that security.

Whatever this upswing in demand is due to, it is certain that the changes witnessed in the intensive market will also be seen in the organic market over the ensuing years so in planning the future of organic poultry meat it has to be accepted that it will not remain a static scene. There may only be a narrow divide between fulfilling the needs of the market and an oversupply, and once an oversupply position is reached the very first thing to be affected is price and in consequence profitability. The organic market is not immune to the same issues that have affected the intensive poultry market, so in entering the realms of organic poultry meat production an element of caution and a considerable degree of thought and understanding are necessary.

In considering poultry meat production within an organic farming operation, it should not be viewed as a standalone enterprise but as part of the whole practice of organic farming. Poultry must contribute to the equilibrium of the farm in terms of adding to the overall profitability, adding to soil fertility and contributing to pest and insect control in a rotational programme. It must form a link in the chain of sustainable agriculture.

7.2   Choice of market

Unless the intention is to supply one's home needs only and not to produce on a semi-commercial or commercial scale, the first decision concerns the market into which the chickens are to be sold. Whatever farming venture is undertaken there must be a clear view as to the end market so that the product meets the precise needs of the consumer. In the case of organic poultry meat this is particularly relevant, considering the high cost of the major inputs and the end price of the finished product. The choice rests between being an individual supplier meeting the needs of farm-gate sales, or farmers' markets, or alternatively linking up with a processor in an integrated manner to supply into either the catering trade or the supermarket outlets. Here it is worth remembering that currently 76% of all chicken sales lies with the top six supermarkets, and the top ten account for nearly 90%, whereas sales through butchers' shops account for 4.8% (British Chicken Information Service 1999).

The production system is different depending on whether one is to continuously produce small numbers on a multi-age basis for the farm-gate market, or batch produce as part of a group of farmers supplying an integrator, who in turn supplies the supermarkets or catering outlets. Each has its merits, so in order to make this early choice it is worthwhile looking at how the two systems operate.

7.2.1 *Farm-gate or local sales*

With this system of production the farmer will have to source a supply of chicks in relatively small numbers on a regular basis depending on the eventual rate of mature chicken sales. Small independent hatcheries exist to meet these demands and advertisements appear in the various trades' journals giving details of supply. It will not be possible to obtain chicks from a local commercial hatchery as the strains will, in all probability, be the hybrid varieties and not the necessary slow-growing strains. In the early stages this is a difficult ordering pattern to establish, but as trade develops and regular orders replace random sales the position becomes easier.

The strain of chicken should be a slow-growing strain capable of reaching 81 days of age prior to slaughter without becoming too heavy and hence too expensive to preclude a sale. It must be remembered that once the unit price of the chicken exceeds that of the customer's budget, the cost of feeding that chicken continues and the live-weight will continue to rise, thus making the chicken yet more expensive. Therefore the ability to market all the chickens at the correct age and weight is one of the key features of a successful business.

Regulations exist which govern the kilograms of live-weight per square metre in the chickenhouse (see Table 7.1). These densities vary depending on the standards of the regulatory body under whose auspices the chickens are being grown, so if the chickens are not marketed at the expected weight the density in the house could increase beyond the regulatory level, and therefore the specification of organic would not be achieved. Linking chick purchase with marketing age, weight, and stocking density is unavoidable, and time and effort must be spent on this aspect.

The option to produce a home mix feed has appeal but the reality is different. So much of the health of any animal depends on the quality, quantity and nutritional balance of the feed that the days of scattering a handful of wheat, organic or not, and allowing the bird to forage for the remainder of its diet have gone. There will always be those who say that they achieve success by this means, but our belief is it is far better to purchase feed from a registered organic feed supplier than to mix a home produced feed. A compounded feed contains a mix of raw materials with documented and traceable supply routes, and a mineral–vitamin balance that will ensure the health of the chickens.

In the view of local authorities it is normally assumed that farm-gate sales only cover unprocessed goods produced on that farm. It is ancillary to the use as a farm and therefore does not require specific planning permission. If the product is labelled or described as 'organic' it must, regardless of the size of the operation, meet all the requirements of UKROFS and Trading Standards.

7.2.2 *Supermarket production*

In producing for supermarkets, many of the decisions are not within the control of the farmer, and neither are many of the worries. This type of organic production may not suit the farmer who wishes to take all the risk, as is the case with many other aspects of farming, however, the market is secure and there should be firm knowledge that all the poultry will be purchased at a pre-agreed price. How therefore does this supply structure operate?

Although integrated companies operate in slightly different ways, it is common practice that they secure the supply of the day-old chicks. These may be provided either 'as hatched', i.e. an equal mix of males and females, or as sexed chicks where the farmer receives either males or females. The date and time of chick delivery is arranged in advance and under normal circumstances a full house, or houses, is delivered at the same time. This allows the house(s) to be ready for occupancy at least two days before chick delivery in order that the correct brooding temperatures and food distribution have been achieved.

The source of food is selected and the manufacturer audited against the supermarket and the regulatory bodies' standards. Food is ordered in advance of the chick delivery so that the food hoppers or track are filled and awaiting the arrival of the chicks.

It is customary that when the chick delivery date is known the slaughter date is also scheduled for 81 days hence. The farmer therefore knows he has no marketing problems. His or her responsibility is to produce a quality product at the weight and food conversion that will achieve optimum returns under the management and welfare standards expected by the final consumer and by the supermarket that puts its name and reputation on the end product.

The supermarket will have selected the choice of production system. Whether the chickens are grown to EU standards, UKROFS standard or to Soil Association standards will have been decided as part of the retail marketing plan, so the farmer will have little or no say in this matter. It is obvious that the more demanding the farming criteria the greater will be the cost of production, and this is reflected in the price paid for the product. In turn the processor must reflect this in the price paid to the farmer for the live organic chickens. If the highest level of production is demanded it will be the retailer's responsibility to inform the customer why the eventual retail price is higher than organic chicken produced to a less demanding agricultural standard.

Different integrators may operate different payment systems. At one end of the scale are those integrators who provide the farmer with chicks, feed and veterinary advice free of charge, and in return pay him a management fee based on performance parameters such as achieving target weight and food conversion. This system has the advantage of removing much of the risk and the need for working capital.

Alternative systems pay the farmers at the end of the crop on a live-weight basis, usually pence per kilo over a weighbridge, and debit from this payment the cost of the feed and chicks. As with the previous system, it does away with the need for working capital, which can be considerable, bearing in mind the cost of the feed, the poor food conversions and the age of kill. For example, to rear 1000 chickens on food at £300 per tonne, food conversion at 3 : 1 and a live-weight of 3 kg results in a food bill alone of £2700. On top of this is the cost of the chicks which will very much depend on the breed selected but will certainly be a further £500–£600.

The integrator is responsible for collecting the chickens on the day of kill and transporting them to their licensed processing plant. When the number of birds is relatively low the farmer may have to catch or assist in catching them on the day of slaughter.

Farming can be a lonely occupation and it should not be overlooked that growing chicken as part of an integrated group can involve more contact between producers of similar minds as well as the ability to share experiences and problems. Most integrators have experienced field staff who routinely visit farms or respond to calls for advice and assistance. This is especially relevant to those farmers who have no real experience of poultry production other than keeping a few laying hens or bantams almost as an ornamental feature. When a farmer embarks on supplying organic chicken to a supermarket the responsibility takes on a whole new meaning; it becomes a major part of the farm economy as well as a vital part of the farm's biodiversity and crop rotation.

Terms such as traceability and auditing are not just words to which lip service is paid. The supermarket and the retail customer expect far more from the producer of organic chicken than that expected of standard production, after all, the product is priced considerably higher than standard chicken and for that reason alone only the most rigorous standards can be considered acceptable in meeting customer expectations.

7.3   Production standards

Whether or not the supermarket or the processor has decided the production standards, or if they are to be decided upon by the farmer, the subject of standards needs to be understood.

They have a significant effect on the whole ethos of production, affecting the land area, stocking density and the capital required to construct the buildings, to name just a few of the main variables. The differences in standards also affect the ultimate cost of the finished product, so the knowledge is necessary to either select the production system that favours the unit cost of the end product into a price sensitive market or to defend the cost if the higher standards have been chosen.

Organic poultry production in the UK is regulated or guided by several sets of standards. All UK producers who wish to have their products labelled as organic must comply with EU standards unless there are derogations, and they must comply totally with the United Kingdom Register of Organic Food Standards (UKROFS) which represent a nationally agreed definition which is legally enforceable.

EU regulations

Underpinning the standards required to produce organic poultry meat lie the European Council Regulations. The first of these was (EEC) No. 2092/91, which came into force on 22 July 1991 and focused solely on crops and crop products. Subsequent legislation has been brought into being which deals with animal production, the latest of which is (EC) No 1804/99, 9104/99 ADD1 and 2. At the time of writing, the agreed text of this amended proposal has yet to be published in the *Official Journal* of the European Communities, however this is a key document as it contains the text of the amendments to the basic Regulation (2092/91).

United Kingdom Register of Organic Food Standards (UKROFS)

In 1987 MAFF recognised the need to establish a body that would bring together the various organic producers and the various standards. UKROFS was therefore to become the certification authority for organically produced food. As part of this role a Poultry Working Group was established that set standards for poultry production. The standards that they set could equal but not fall below those standards required by the EU. Article 12 of 9104/99 allows member states to apply more stringent rules than those laid down by the Council Regulation for livestock and livestock products produced within their own territory, however trade restrictions cannot be applied to livestock produced elsewhere in the Community.

It can be seen therefore that a two-tier set of standards exists, namely the EU standards and UKROFS. Poultry meat produced and sold in the UK has to meet UKROFS requirements, but organic poultry produced in mainland Europe can be exported into the UK unhindered, provided it meets the EU standards, both carrying an organic label. This would present no problem in production terms unless the UKROFS standard were more stringent than the EU, when the home produced poultry would be at a disadvantage faced with cheaper imports.

The issue of standards is further complicated inasmuch as within mainland UK there are three organisations that can certify organic production under UKROFS, each with their own requirements and standards. They are:

(1)   The Soil Association, Bristol House, 40–56 Victoria St, Bristol, Avon, BS1 6BY.
(2)   Organic Farmers and Growers, 50 High St, Soham, Ely, Cambridgeshire, CB7 5HF.

(3) Organic Food Federation, The Tithe House, Peaseland Green, Elsing, East Dereham, Norfolk, NR20 3DY.

These private bodies, approved by UKROFS, have for their own reasons elected to adopt differing requirements, some of which are stricter than those of UKROFS. This further complicates the choice that the producer has to make, not only in marketing terms but also in the level of investment. If this is confusing to poultry producers, how much more so it must be for the consumer, unless good reasons are offered through quality-marketing, enabling customers to make an informed choice. Ultimately there must be a coming together of standards in the interests of customer understanding.

Table 7.1 summarises some major differences in standards. It is not fully comprehensive and is intended as a general guide only. Once the production standard under which the birds will be produced has been agreed, it is important to scrutinise the full text of the chosen standard.

Interplay exists between some of the variables; for example, if the live-weight that is required to be produced is heavy, the first limiting factor will be the kg/m^2 inside the house and not the allowable number of birds within the house. Similarly, on a small farm, if the farmer cannot, or does not wish to, export the end-of-crop litter from the site, the number of birds allowed will be restricted by the limiting factor that the application of nitrogen must not exceed 170 kg per year. (This is the amount produced annually by 285.7 table chickens.)

There are many other EU and UK regulations relating to poultry production which have to be complied with, as well as the specific regulations pertaining to organic production, including:

- Welfare of Livestock Regulations 1994 (SI 1994 No. 2126) although these will be replaced shortly with new regulations to implement EU Directive 98/58
- Welfare of Animals (Slaughter and Killing) Regulations 1995 (SI 1995 No.731)
- Welfare of Animals (Transport) Order 1997 (SI 1997 No. 1480)

The Farm Animal Welfare Council produces recommendations for the welfare of poultry and MAFF produces numerous publications dealing with management and welfare, available free of charge from MAFF Publications. As this is a developing market, the scene can change rapidly, so it is always advisable to seek up-to-date advice from UKROFS or from other bodies if their requirements exceed those of UKROFS.

7.4 Conversion periods

If the farm is already registered as an organic farm there is nothing other than compliance with planning law to prevent the development of a poultry facility. Planning consent may not be necessary if the house is mobile and therefore does not have foundations or a fixed water supply, but it is always advisable to check this with the local planning authority before proceeding. Conversion of land for poultry production involves the same process as for other forms of organic farming, but there is derogation that the conversion period may be reduced for pasture or open-air runs used by non-herbivore species. This period may be further reduced where it can be proved that the land has not, in the recent past, received

Table 7.1 EU Standards, UKROFS Standards and Soil Association Standards for the keeping of organic poultry.

	EU Standards for fixed houses	EU Standards for mobile houses (not exceeding 150 m²)	UKROFS Standards	Soil Association Standards
Breed	Table bird to be of a strain recognised as being slow-growing if they are to be killed younger than 81 days. Preference for indigenous breeds	Table bird to be of a strain recognised as being slow-growing if they are to be killed younger than 81 days. Preference for indigenous breeds	Breeds appropriate to finish in 81 days without welfare implications. Recognised slow-growing not essential	Recommend hardy, disease resistant and slow-growing breeds. Breeds from Hubbard/ISA or Rhode Island Red males or other indigenous strains best suited. Fattened slowly, generally 70–81 days
Chicks	Less than 3 days old when arriving at the farm	Less than 3 days old when arriving at the farm	May be brought in from conventional sources at 1 day old. Must be less than 3 days old	1 day old
Feed	May comprise up to 30% in-conversion feedstuff. Must contain at least 65% cereals in the fattening stage. 80% from organic sources	May comprise up to 30% in-conversion feedstuff. Must contain at least 65% cereals in the fattening stage. 80% from organic sources	Planned that 100% of feed is to UKROFS Standards, where not possible at 80%	Target 100%, but where not available at 80% of dry matter. Minimum 65% cereals
Vitamins	Synthetic vitamins identical to natural are permitted under vet supervision	Synthetic vitamins identical to natural are permitted under vet supervision	Do not accept synthetic vitamins	Preference for natural sources. Concentrated vitamins only when necessary to satisfy normal requirements
Pure amino acids	Essential amino acids allowed but preferably from materials occurring naturally	Essential amino acids allowed but preferably from materials occurring naturally	Do not accept synthetic amino acids	Preference for natural sources. Concentrated amino acids only when necessary to satisfy normal requirements
Vaccination	Allowed	Allowed	Allowed	Allowed if known disease risk
Medication	1 dose medication in life cycle, double withdrawal period	1 dose medication in life cycle, double withdrawal period	May be used if organic treatments fail, only under responsibility of veterinary surgeon. Double withdrawal period	Coccidiostat allowed in starter rations, Amprolium if problem

General points on feed	Normally produced on holding	Normally produced on holding	Land-based integration desirable. Recognise need to bring in feed	Aim to produce feed from holding
Housing: general	Must meet biological and ethological needs. Must be reared in open range conditions, no cages	Must meet biological and ethological needs. Must be reared in open range conditions, no cages	Sufficient ventilation. Dry rest areas. No limit on number of houses per site	Permanent housing permitted but mobile houses preferred. Max 25% of floor slatted; littered area to be dry and friable. Organic and non-organic chickens cannot be reared on same holding
Ventilation	Plentiful natural ventilation sufficient to prevent harmful build-up	Plentiful natural ventilation sufficient to prevent harmful build-up	Sufficient ventilation	Adequate natural ventilation
Lighting	Maximum 16h. Including natural daylight. No artificial light for at least 8h	Maximum 16h. Including natural daylight. No artificial light for at least 8h	As EU, suggest min of 10 lux	Adequate natural lighting. Artificial lighting to a max. of 16h total light
No. of birds/m^2	10	16 only in the case of mobile houses <150 m^2, open at night	12	12
Liveweight/m^2	Max. 21 kg	Max. 30 kg	Up to 25 kg	Recommended 15 or 18 kg, permitted 25 kg in mobile houses, 21 kg in fixed
House capacity	4800 birds max.	4800 birds max.	Stable colonies. 2000 max. pending research	Recommended 200 (permitted 500)
Pop holes	4 m in length/100 m^2 of floor area	4 m in length/100 m^2 of floor area	4 m in length/100 m^2 of floor area	Continuous and easy access to runs
Outside drinkers	Yes	Yes	Yes	Yes
Feeders	Outside feeders required	Outside feeders required	FAWC and EU welfare standards. No outside feeders	

(*Continued.*)

Table 7.1 (*Continued.*)

	EU Standards for fixed houses	EU Standards for mobile houses (not exceeding 150 m²)	UKROFS Standards	Soil Association Standards
Houses per production unit	Usable house area must not exceed 1600 m²	Usable house area must not exceed 1600 m²	No limitation	
Bedding	At least one-third of floor solid. Litter straw, wood shavings, sand, turf	At least one-third of floor solid. Litter straw, wood shavings, sand, turf	At least 75% of floor solid. Litter straw or other appropriate material. All animals have dry lying areas	Recommended: straw from organic sources. Permitted: non-organic straw and untreated wood shavings
Between crops	Houses cleaned and disinfected, vegetation to grow back			Houses disinfected, methods listed
Outside: range area	4 m² per bird, or 2500/ha provided the 170 kg N/ha per year is not exceeded	2.5 m² per bird, or 2500/ha provided the 170 kg N/ha per year is not exceeded	Recommended: 10 m² per bird, or 1000/ha	625 birds/ha, i.e. 250/acre or 16.2 m² per bird. Increased if pasture is rotated. Pasture to be rested 1 year in 3
Access to range	At least one-third of life and whenever weather conditions permit	At least one-third of life and whenever weather conditions permit	Access from 6 weeks. Protect from predators	May be brooded inside for 28 days. Access to outside for two-thirds of life
Range quality	Mainly covered with vegetation and with protective facilities. Specified rest periods	Mainly covered with vegetation and with protective facilities. Specified rest periods	Trees and shelter to provide shade and weather protection. Appropriate rest periods	Access to suitable shelter. Rested 1 year in 3 if set stocked
Distance to range	No recommendation	No recommendation	Currently no standard but a recommendation that max. ranging distance is 250 m from house	
Manure utilisation	Not to exceed 170 kg N/ha per year, i.e. 580 table chickens per ha	Not to exceed 170 kg N/ha per year, i.e. 580 table chickens per ha	As EU but accept excess manure could be exported to other organic unit	As EU but accept excess manure could be exported to other organic unit
Processing: journey times to slaughter	Subject to Community or national rules or recommendations	Subject to Community or national rules or recommendations	Max. 10 h including loading but would like 6 h	Max. 8 h including loading

treatments with products not approved by the organic movement. This derogation must be authorised by the inspection authority or body.

7.5   Breed selection

The choice of breed or strain must take into account the capacity of the birds to adapt to local conditions, their vitality, and their resistance to disease. In addition, breeds or strains should be selected to avoid specific diseases or health problems, which may be associated with some strains developed for intensive production.

In an integrated production system the choice or selection of breed may not be a decision which the farmer has to take inasmuch as the processor or the retailer may have already made that choice. The independent farmer however is in a different position so an understanding of the choices is important, and likewise the integrated farmer is better informed if the reasons for selection are understood.

One standard that all organic farmers have to comply with is the 81-day minimum age at slaughter, although the EU regulations say the minimum age at slaughter shall be 81 days for chickens *unless* they are a slow-growing breed, when presumably the killing age is undefined. This appears to be a contradiction in terms for which no explanation is available. The EU proposes that a standing committee prepare a list of slow-growing breeds for the purpose of the regulations.

When the age at kill, or, in the case of the EU regulations, the definition of a slow-growing breed, is linked to the live-weight of the bird at kill, a major influencing factor in breed selection is apparent. The standard Cobb, Ross and Hybro breeds have been developed for the intensive market where weight for age is a necessary requirement and therefore by definition must fall outside the parameter of a slow-growing breed. To expect these strains to reach 81 days yet still be in the 4.5–5.5 lb live-weight band, giving a carcass weight of 3.25–4.0 lb, is too much to hope for, even allowing for very low-density diets. To impose feed restrictions that would be necessary to even approach these live-weights would be both severe and against the ethos of organic production.

If the live-weight for the market into which the birds are to be supplied is significantly higher it may be possible to rear only the females of these breeds to the required age of 81 days, but it needs to be remembered that the unit price of such a product may be outside the economic expectations of the average consumer. This is less of an issue if the meat is to be sold as portions when the unit cost can be broken down into smaller parts, but here again the relative value of the white meat against dark meat has to be considered.

If it is not possible to use the readily available commercial breeds the farmer may be able to look for the speciality breeds supplied by companies such as Hubbard/ISA who have developed several breeds of chicken especially for the specialised market. Examples of strains from Hubbard/ISA are the Redbro, which is already used in the free-range and traditional free-range markets in the UK, and the Red JA. With these breeds live-weight expectations are lower. The published figures in the Hubbard/ISA manual refer to an 'as hatched' population of Redbro achieving 2.209 kg at 56 days of age compared with a bird bred by Cobb or Ross weighing in the region of 3.3 kg at the same age.

No breedstock company projects the live-weight of its birds to 81 days of age, which is required by the organic standards. If we project a live-weight gain of between 50 and 75 g per day, and for the sake of the calculation we use 58 g per day, we arrive at an 81-day weight of 3.65 kg. Likewise the Cobb or Ross bird would achieve 4.75 kg. The carcass weight of the respective breeds will be 2.6 kg (5.8 lb) and 3.4 kg (7.5 lb). The high unit cost of producing organic poultry means that it is essential to control weight gain by not feeding *ad lib*, even low-density diets. With all breeds there are considerable weight differences between males and females, so depending on the weight ranges required by the customer, the need to sex grow, i.e. grow either males or females, will be established. This will allow a narrower weight spread to meet market needs.

Part of the breed/live-weight decision includes whether to select a breed or strain with white or yellow skin, or even with black or white hocks. In the UK, customers are used to having white skin chicken whereas in the US yellow is the norm. The customer's attitude to changing this trend needs to be considered. It may be that the appearance of yellow skin and fat is counted an added value, in which case rearing strains of chicken with this trait will be a distinct advantage, but on the other hand in these times of 'healthy eating' the presence of yellow fat can be off-putting. And when the consumer is only used to seeing white or yellow hocks, what is the reaction to change? Hence the market into which the birds are to be sold needs to be researched to decide the choice of breed.

To use native or traditional breeds presents difficulties if any significant level of volume or regular supply is to be achieved. The establishment of parent flocks providing fertile hatching eggs on a regular supply basis will be required. These parent flocks will have to meet the necessary MAFF testing and inspection schedules. Traditional breeds of meat-type strains may prove acceptable for small-scale production by the self-supply producer, but otherwise impose severe limitations.

The choice of breed is not an easy decision and care taken at an early stage of project planning can influence the outcome of the organic venture.

7.6   Feed selection

If the choice is to purchase feed from an approved organic feed mill the factors to consider are availability, quality, traceability, and continuity of supply and price. A major difficulty facing certified mills is the supply of approved organic raw materials. Such is the shortfall in cereals and vegetable proteins that manufacturers have to import from Europe and as far east as Hungary and Romania. All that is required is a certificate to say that the product falls within the necessary organic parameters. No test can be applied which can distinguish between organic and non-organic, and the farmer paying high premiums for organic feed is placing reliance on the integrity of the compounder.

Within the UK the position is further exacerbated by the shortage of land for organic cereal production, not helped by the current shortage of funds available under the government's organic farming scheme.

Advice on the composition and formulation of the feed is best left to the nutritionist. A reliable company will provide records of raw material traceability and assurance of compliance with the correct permitted percentage of non-organic fraction.

7.7   Location and land selection

In locating the houses within the farm it must be remembered that access for chick and food deliveries will require hard roads, unless they are to be unloaded at a central point for onward movement by an all-terrain vehicle. The collection of the grown birds also needs resolution before the arrival of the collection lorry if an off-farm slaughtering facility is to be used (see section 7.15).

Land on which range birds are kept for any length of time may become 'chicken sick', which is a convenient term for stale land heavily contaminated with droppings and consequently heavily laden with bacteria and other disease-causing micro-organisms prejudicial to the health of the stock. The time taken for this to happen depends on the nature of the soil, wet and poor draining land being worse than light or chalky soil. It also depends on stocking density, the ability of the birds to range away from the house and whether or not the house is mobile. The author's belief is that all organic houses must be mobile and moved to a new location after every crop. To be easily moveable the house should not be larger than that needed to house 500 birds, which is 31.25 m^2 or 336 ft^2. To describe a house as mobile and not to move it is certainly not giving the consumer what they expect!

Where fixed housing is used, the EU regulations reduce the interior stocking density from 16 birds to 10 birds per m^2 and the live-weight from a maximum of 30 kg/m^2 to 21 kg/m^2, and a programme of paddock rotation put in place so that one paddock or area is resting while three others are in use.

Whichever housing system is employed, the grazing of the pasture with another species (sheep, for example) is a good means of keeping the grass short, allowing sunlight to destroy parasites.

Regulations may establish the maximum number of birds kept on the land, but good farming practice will establish whether these maximum levels can realistically be maintained. These figures may need to be adjusted downwards if soil type or local rainfall levels are adverse, to preserve the integrity of the soil and to maintain bird welfare.

7.8   House selection

The differences in standards between UKROFS and, for example, the Soil Association are apparent when it comes to purchasing or building the necessary poultry house. UKROFS allow for an undefined colony size described as a 'stable colony', and recommend (UKROFS Recommendations Regarding EU Poultry Proposals: Final Version 30 November 1998) that the standards should be at least those of the 'traditional free range' defined in EU marketing regulations, which refers to a colony size of 4800 birds. In contrast, the Soil Association recommends colonies of 200 birds but will allow 500 birds. This creates a fundamental difference in the house dimensions and hence the cost per sq.ft. of the house itself.

If the decision is made to produce organic poultry in the small 500-bird mobile unit, the choice between skid or wheeled units arises. Whichever is chosen, the need for a tractor of sufficient size will be necessary, bearing in mind the need to move houses under winter conditions. It is not a fast movement and at least an hour should be allocated to simply move from one land area to the next.

7.9   Power supply

Power to houses can be achieved by 240 V mains or 12 V battery, charged with a wind powered unit, which in turn can be backed up by either an integral or a remote battery charger. Power requirements for a 500-bird mobile house are not high, especially if suspended tube feeders are used which are hand-filled from bagged deliveries of food. Automatic feeding systems are available which can utilise 12 V if mains power is not readily available. Lighting can be from natural light via a roof window and this could be supplemented with a 240 V or 12 V system if needed. Poultry meat birds do not have the need for controlled day length, as do laying hens.

7.10   Planning permission

It will always pay to consult with the local planning authority. Advice is given in the local authorities' Planning Policy Guidance Note No.7 which in paragraph C11 states:

> 'The Courts have held that some temporary structures used for agriculture are not "buildings" in planning terms but are a use of land and so outside the scope of planning control. Thus, temporary accommodation for livestock, such as "pig arks" and moveable poultry shelters may not be "buildings" for planning purposes. The status of a particular structure is ultimately a matter for the Courts to decide, on the facts of each case. A structure placed on foundations, secured to the ground and with, for example, facilities such as an integral water supply may constitute a building, whilst a structure without such features may constitute a use of land. In case of doubt an application may be made to the local planning authority for a certificate of lawfulness of proposed use or development under section 192 of the Town and Country Planning Act 1990 (as amended).'

7.11   Chick quality and brooding

Regardless of whether the chicks are home produced or bought in as 'day-olds', the quality of the chick can determine the success of the flock. This is particularly the case with organic production where antibiotic support for common yolk sac infections would not be considered appropriate in other than extreme circumstances.

The chick, by virtue of the retained yolk, can survive for 3–4 days without food or water. This should never be a survival mechanism that is put to the test, but it does explain why at 3–4 days of age it is common to see an increased mortality. Chicks with low vitality, and especially those from young parent flocks, often appear to be active until this age and then account for a mortality increase. As a guide the first week's mortality should not exceed 1% and ideally be no higher than 0.75%. The time spent obtaining ideal conditions for the chick in terms of temperature, food, water and freedom from draughts cannot be overstated, but in a chapter such as this it is not possible to cover the finer points of management practices. It is sufficient to say that the brooding temperature of 30–32°C can be achieved by LPG (low-pressure gas) canopy brooders, by infra-red lamps, or by heat mats similar to those used in pig creeps. Whichever system is selected, the heat source must be controlled by a thermostat and the house must be alarmed for both high and low temperatures.

If the farmer has little first-hand experience of poultry husbandry, these points should be learnt from the management advisor employed by most integrators or from specialist textbooks or manuals produced by the companies supplying stock into the intensive market. However it must be remembered that when the inexperienced producer is only learning from textbooks or, even worse, learning from the ill informed local 'expert', the welfare of the stock is in jeopardy.

7.12 Veterinary advice and poultry health

Before starting organic production it is essential to identify a veterinary practice which can offer sound advice with regard to poultry and which has a good understanding of the needs of organic production. Being a good cattle vet is not necessarily the same as being an expert in poultry diseases and their management, especially when choice of vaccines and antibiotic medication for the ailing flock is restricted or not permitted.

So much of disease prevention can be based on the selection of the appropriate breed, good nutrition, husbandry and environment. The avoidance of overstocking linked to controlled pasture rotation goes a long way to control many of the 'diseases of intensification'. This is well summarised by the Farm Animal Welfare Council as 'positive welfare'.

Should a form of treatment be required beyond the scope of positive welfare, the EU regulations allow the use of phytotherapeutics, e.g. plant extracts, essences, and so on, but exclude antibiotics. Homoeopathic products of plant, animal or mineral origin are permitted and also certain trace elements that are categorised in Annex II of the regulation. Should the approved products fail to be effective in controlling or curing the problem, and other forms of treatment are necessary to avoid suffering or distress to the poultry flock, then conventional medication can be used, which may take the form of chemically synthesised allopathic products or antibiotics under the control of the veterinary surgeon retained by the producer. Under EU regulations the use of these products is not permitted as a form of preventive medication and neither are those products designated 'growth promoters' or 'antibiotic digestive enhancers'. Under UKROFS, preventive chemotherapy may be used to deal with specifically identified diseases or as part of an agreed conversion or disease reduction plan.

In the UK all veterinary medicines must be used in accordance with their UK product licence or as directed by the prescribing veterinary surgeon. Withdrawal times from the last day of treatment to the day of slaughter shall be at least double that time laid down in the product licence and shall not be less than 14 days in the case of controlled drugs or prescription-only medication (POM). The EU regulations stipulate twice the legal withdrawal period or, in the case of it not being specified, a 48-hour withdrawal.

Treatment with organophosphorus compounds excludes chicken from being sold as an organic product. Under the EU regulation, with the exception of vaccinations or any compulsory eradication schemes established by member states, a flock of poultry receiving one course of treatment with a chemically synthesised allopathic product or antibiotic may not be sold as organic.

In the UK vaccination is permitted where there is a known disease risk. Vaccine choice and programming should be agreed with the veterinary surgeon to ensure adequate disease protection and where possible a progressive reduction in use as the organic unit becomes established.

For health reasons buildings must be emptied at the end of the crop, cleaned, washed and disinfected. In the EU regulations, a list of approved cleaning and disinfectants is given, including hydrogen peroxide and formaldehyde, both of which have been proven to be valuable products in the intensive sector. Regarding other aspects of disease control, it is advisable to operate an 'all in–all out' policy on the site rather than to operate a multi-age site where diseases can be transmitted from adult birds to incoming chicks. Poor performance or high mortality will rapidly reduce profits and negatively affect the welfare of the stock.

7.13 Welfare

One of the keystones of organic production is welfare, and in providing the necessary free range this has to be taken into account. The same governmental controls exist for organic farms as for the intensive sector. The presence of disease, poor land management, poor litter, and low temperatures linked to frozen water supplies all impinge on bird welfare. It must be remembered that in the mind of the consumer the organic label implies the very best of welfare.

MAFF publishes useful codes of practice on heat stress, litter conditions and other management aspects (see Table 7.2).

7.14 Predators

Land predators can be deterred by wire fences 7 or 9 strands high, with an outer electric fence wire sourced from a mains conversion unit or a battery.

Table 7.2 Publications available from MAFF.*

Title	Reference
Heat Stress in Poultry	PB 1315
Farm Fires	PB 0621
Codes of Recommendations for the welfare of livestock standards:	
Domestic Fowls	PB 0076
Poultry Litter Management	PB 1739
The Water Code	PB 0587
The Soil Code	PB 0617
The Air Code	PB 0618
Welfare of Livestock Regulations 1994	MAFF
Welfare of Animals (Transport) Order 1997	MAFF
The Welfare of Poultry at Slaughter: A Pocket Guide	PB 3476

*MAFF Publications, Admail 6000, London SW1A 2XX, Tel.: 0645 556000.

Predators from the air pose a more challenging problem. Birds at free range may be ready to go outside for limited periods as early as 10–12 days old in the summer and at this age are easy prey to magpies, rooks, buzzards and so on. There are various methods of protection but none which are absolute, so inevitably some young birds will be taken, one of the less appealing aspects of bird welfare at range. Dummy owls and so on have been tried, as have distress call recordings, but in the author's experience do not meet with much success; they also may have an adverse effect on the chickens that are being encouraged to range. Interrupted flight paths and good cover by shelter screens or even old trailers parked in the range area give the chickens some protection in the early days. Tree planting within the range area is to be encouraged but this can allow perching sites for some of the other avian predators.

7.15 Collection and transport

A conventional four-wheel drive vehicle with a 5 m trailer is adequate for collecting around 500 birds, while a more custom-planned collection service will be necessary for over 750 birds. Concrete or hardcore roadways may service fixed houses, but the mobile units in winter may only be accessible by farm tractors. These may have to ferry the catching modules to and from the waiting collection lorry parked at a convenient location where a fork lift can transfer the full module from the farm trailer to the lorry and reload an empty module to the farm tractor and trailer. All of this requires good communication and co-ordination between the farm, the catching units and the processor in order to get the chickens into the processing plant with the minimum of delay and stress to the birds.

As the majority of chickens will be killed in a conventional processing plant approved by the certification body, the organic birds will be the first batch of the day to be processed. This means that the birds will need to be caught in the early hours of the morning while it is still dark, which happens to be a good time to catch flighty birds.

References

British Chicken Information Service (1999) *British Chicken Marketing Review 1999*, British Chicken Information Service, London SW7 4ET.
Hubbard/ISA, Warren Hall, Broughton, Chester CH4 0EW.

8 Organic Dairy in the United States

Louise Hemsted

Organic dairy in the United States has its roots and history steeped in conventional dairy production. The United States is different from Europe due to the extensive logistical infrastructure, which allows product to be consumed literally thousands of miles from where it originated.

8.1 Conventional dairy issues

Bulk tankers of conventional milk routinely move from the midwestern states to those areas and climates that are not as suitable for dairy production. The methodology for payment of conventional dairy farmers, known as the Federal Milk Marketing Order system, includes an adjustment based on the distance the farm is located from 'ground zero', which is in Eau Claire, Wisconsin, with the lowest prices being paid to farmers in the midwest. The Federal Order was originally introduced to stabilise the marketplace for different grades of milk. The effect is to tax Grade A use of milk (fluid milk and 'spoonables' such as yogurt and cottage cheese) and pay back to Grade B dairymen (those producing milk destined to become cheese, butter and dried dairy products).

The same system of payment is applied to organic dairy farming. However, all organic farmers are required to be Grade A producers, and receive the same payment, whether the milk becomes cheese or is sold as fluid. Therefore the Federal Order system does not perform a balancing function between Grade A and Grade B products for the organic farmer, and is effectively a tax on organic dairy farming.

Midwestern farmers are selling and dissolving multi-generational farms at an unprecedented rate, which has led to the federal order system frequently being challenged in recent years; nonetheless it remains in effect.

The prices received by the conventional farmer have little to do with the actual cost of production, which is based on land values, availability and location of feed, and bovine production. In recent years, however, we have seen the migration of the dairy farms to 'less profitable' regions because of the higher premium paid for conventional milk. In some cases this has been successful and in other cases not. We find ourselves irrigating desert land and pumping water from aquifers deep below the surface. It is not in the opinion of this writer sustainable or profitable.

Organic dairy has many faces and models in likeness to conventional dairy. The original model is family farm based, and I believe the more sustainable method. Family farmers who grow at least 50% of their feed, and are active in the day-to-day management of the farm appear to have good success. By trading off their physical labour, they return more dollars to the acre than the conventional farmer. The second model is the 'modern' model where cattle are kept en masse, and feed is brought to them. This model has many of the problems and environmental issues (manure/waste management) that plague the conventional farm. However, it lends itself to efficiencies in freight of the milk from the farm to the processing locations.

Organic dairy officially evolved in the late 1980s – based on farmers/producers around the nation who had rebelled against the continual forces of pesticide application, herbicide application, antibiotics, and hormones.

8.2　Organic dairy pioneers

The Strauss Family Farm of Marshall, California, was perhaps the first in the market with organic milk bottled in glass. This family farm felt it was in its best interest to produce and bottle its own milk, taking it to market and minimizing the middle distribution stream. It was and continues to be successful, having added butter, cheese and yogurt to its distribution. The company also purchases organic milk from other farmers to support its brand. CROPP – now known as Organic Valley – emerged in the Midwest in 1988, first as an organic vegetable co-operative, then dairy, eggs, and subsequently meat. Its origin was a group of seven dairy farmers, who established a pay price based on what they felt it cost to produce organic milk and return a small profit to the farm. In the beginning their actual pay price was the average between what they sold on the organic market and what they sold on the conventional market. In the north-eastern United States, Peter and Bunny Flint organised farmers, bought their milk, and developed a brand known as Organic Cow.

These three were the originals, with CROPP being the first to pioneer national and international distribution. In 1993, Horizon Organic Dairy began purchasing milk from CROPP, and eventually built an 800-cow dairy herd, now expanded to 3000 cows in Idaho. Horizon was joined in the market by various other start-up organisations, but none has yet developed the national distribution that the leaders CROPP/Organic Valley and Horizon have developed.

8.3　Farms, family and corporate

8.3.1 *Family farm model*

The family farm model is based on family farmers, scattered across the landscape. Each producer must comply with Grade A and Interstate Milk Shippers (IMS) requirements, and have the capacity to store four milkings of milk between pickups. The milk is loaded on insulated tanker trucks, and taken to a plant within a 100-mile radius of the farm. At the plant it is segregated, and produced into a variety of organic products. The plant/facility also must undergo certification. The product is then aged, staged, consolidated, and subsequently shipped to a consolidation warehouse across the country for distribution and sale (see Fig. 8.1).

Family farm model – economic biases

Two of the strengths of the family farm model are the local production of feed and the local production of dairy products. Both of these fuel energy and strength into the local

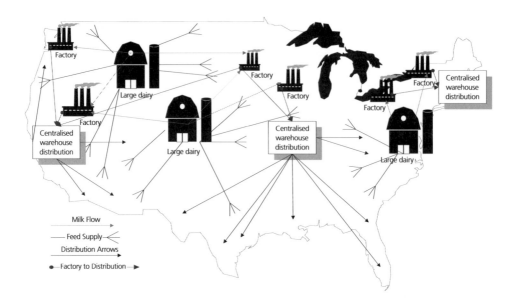

Fig. 8.1 Family farm-based organic dairy production and distribution.

community. When farmers prosper, they spend money and fuel the local economy, which in turn strengthens the regional economy. Economic concerns support family farm agriculture. The recommended ratio of personnel to cows is approximately one person per 50 animals. In family farm agriculture this roughly translates as 20 families per 1000-cow 'modern' dairy, that is, 20 families contributing in different communities picked up by one or two different milk routes (thus supporting two more families), delivering to cheese plants for processing, and on to distribution and marketing.

These families all have the opportunity as small businesses to grow and further develop their enterprises. They are their own boss. Solid organic dairy pricing allows the farmer to plan on his expansion, or lack thereof. Commitments to milk haulers based on a mileage payment allows them to budget and plan constructively. Farmers now constitute less than 3% of the population in the United States, however much of the nation's economy is driven by the wealth of the farmer. When farmers prosper, the economy is healthy.

Product is transported via tankers to the local dairy, and converted into cheese, butter, powder or fluid milk prior to making the longer trip to the distribution centre. While there is a cost to small farm milk pickup (around $15–25 per farmer, on average $1.10 per hundredweight of milk), the objective is to keep the milk in a fairly localised area. This reduces the cost of the final product by transporting the milk in a more concentrated form (such as butter, powder or cheese) on to its final destination.

8.3.2 *Modern (corporate) farm model*

The 'modern' model generally relies on feed brought in from surrounding areas because the large number of animals housed in one location cannot be supported by the available local feed. This model does support family farms, in that they often purchase their feeds from small farmers. The dairy itself must meet GRADE A and IMS requirements.

One discernable difference is that the milk is often hauled hundreds of miles to the manufacturing facility, and then further transferred in its final form to distribution facilities (see Fig. 8.2).

Modern farm model – economic biases

Using the same assumptions discussed above, again, a 1000-cow dairy requires 20 employees. However, above the wages of the 20 employees there is a required payback to the business owners. Given that revenues are the same for the milk production, the piece of the pie is normally divided to return more to the owners than to the employees. This does not mean that the employees are not supported, but that they have limited opportunities for individual or business growth. Their working arrangements do not allow them to plan on methods of increasing their revenues, therefore decreasing their economic inputs into the community.

From a milk hauling perspective, contractors without local ties are often used to haul milk upwards of 500 miles to a manufacturing facility. This also does not support the local economy, except for the regular purchase of fossil fuels.

In addition to the economic concerns, there are concerns about animal welfare in large dairies. A movement is currently under way to require pasture for all dairy animals in various stages of production, in part to spare animals from spending their entire life on concrete.

8.4   Sustainability of organic agriculture

Dairy animals contribute a significant source of phosphorus minerals through their manure that is returned to the soil, from which feed is harvested. In the ideal setting, an organic farmer grows his own feed, purchases a few supplemental vitamins, and feeds his cows, which provide fertiliser for the next year's growth. The addition of animal manure, combined with careful land management, such as rotating corn with alfalfa (alfalfa naturally adds nitrogen to the earth as green 'plow down') creates healthy soils.

8.4.1 *Crop management*

Crop management is complex as each field is different from its neighbour. It is important to test and monitor the soils, so the farmer is conscious of nutrient balance. Since synthetics are not applied to compensate for mismanagement, a field abused can take 6 to 7 years to recover.

As a general rule, corn is grown as the main source of energy in the ration. However it is best to not grow corn on a field for two consecutive years. This often results in depressed yields, as there is not enough nitrogen fixed in the soil to give the corn a boost prior to the emergence of weed growth. By rotating the crop, the farmer also interrupts the cycle of common pests such as the corn borer. Common methods for controlling weeds in corn are the rotary hoe, cultivation, and cultivation with propane burning.

Soybeans are a common source of protein. They also are hard on the soils and should not be grown in consecutive years on the same field. Alfalfa is a good rotational crop for corn or beans. In some areas of the United States, alfalfa does not grow easily. In those

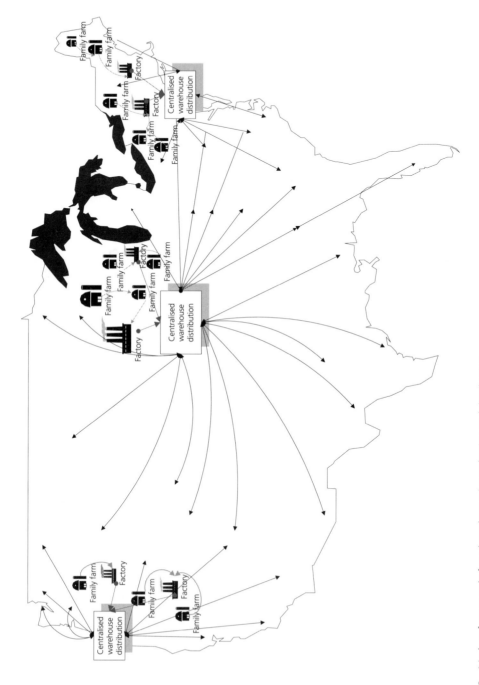

Fig. 8.2 Modern factory style farm–based organic production and distribution.

regions, the primary source of hay is a variety of clovers and grasses. These, with alfalfa, comprise the majority of the roughage the cow will consume. Ensiled at 1/10th bloom, alfalfa silage will contain 22% protein; if baled at the same time it yields slightly less, with some leaching due to environmental exposure.

8.4.2 *Ration balancing*

From the field to the barn, the feeds grown must be harvested in a timely fashion to bring good value to the milking herd. No supplemental preservatives are allowed. Key to any farmer's dairy production is the ration fed to the cow. Typically 90% of organic farmers feed below the recommended level for protein. They meet or exceed the level of energy required for the cow. This simple shift in balance actually improves herd health, while causing only a modest drop in production, therefore decreasing the cost of production.

For farmers purchasing organic feed, the single most important concern is to find a reliable continual source, as cattle do not respond well in production to ration changes.

8.4.3 *Herd health*

In the United States the majority of certification organisations prohibit the use of antibiotics, pesticides and hormones in the dairy animal. Herd health is managed by careful ration balancing, maintaining a clean and dry environment, and careful attention to the animals. Thankfully, due to much public hue and cry about antibiotic resistant bacteria, today's veterinary surgeon is once again becoming practised in treating animals holistically instead of symptomatically. In many areas rich in organic production, veterinary surgeons are dedicating their practices to treating animals in a homoeopathic manner.

Since antibiotics are prohibited, the attention to a clean and dry environment for the animal is critical. The key is the ounce of prevention. If a cow lies in a wet, manure-laden stall every day, the chance of bacterial introduction to the udder increases 10-fold. However this is easily changed in most cases, and becomes the strongest asset to a clean herd. If a cow becomes infected with mastitis, there is a choice of many isolated natural whey antibodies, which when injected boost her immune system to fight the infection. Vitamin C is also widely used on animals suffering from infection. Antibiotics are used in extreme cases, and the animal cannot return to the organic production line, although her life is spared. In many cases the producers sell the animal to a neighbour or to a local stockyard.

Pesticide control is individually based. For flies, fans, sticky traps, ultraviolet lights and pheromone traps are the primary methods of reduction. For lice, clipping the animals and removing the clipped hair with an industrial vacuum cleaner works well.

Hormones are not allowed in organic production. This includes the range from the common milk letdown hormone (oxytocin) to lutylase which is used in conventional dairy to induce egg production by the ovary. Also prohibited of course is the use of the controversial growth hormone known as rBGH (recombinant bovine growth hormone).

Verification of herd health methods, as well as field work, is done annually by an independent certification agency. The inspector will stay at a farm anywhere from several hours to several days, depending on the size, and audit the operation from bookwork to fieldwork. In this process the inspector is on hand for milking, he observes the crops in the growing season, and he audits the farm plan, herd health records and the checkbook.

Through these checks and balances, the inspector develops a report that is then reviewed by an independent committee. These checks and balances are critical to the industry.

8.5   Organic dairy products

The organic dairy sector has been extremely resourceful in creating products that are organic and pleasing to the public. One can purchase organic milk in cartons, glass bottles, and plastic jugs; one can find cream cheese, sour cream, cottage cheese, and hard cheeses from Cheddar to Mozzarella to Romano and Feta. Yogurt of more then a dozen varieties, and ice cream, complete the picture (see Fig. 8.3). Despite all this production, there are only two plants in the United States dedicated 100% to organic production.

Strauss Family Farms built the first dairy plant in Marshall, California. Here they make butter, and bottle milk in glass bottles. Organic Valley built the second plant in Chaseburg, Wisconsin, where they manufacture their European-style cultured butter and reload milk. These two facilities account for less than 10% of the organic dairy production in the United States. The remainder is done by a complex infrastructure of conventional dairy plants.

In the United States, there is a federal standard for fluid dairy production that is known as the Interstate Milk Standard. This is a standard that is supported by the Federal Department of Agriculture (FDA) and United States Department of Agriculture (USDA), both of which carry out various parts of the inspection process. The standard requires that dairy facilities can audit the milk from the farm through the plant and as finished product. Because this infrastructure already exists, the auditing of organic dairy production is quite simple.

All organic milk must be received in through a clean pump, into a clean silo or horizontal tank, with no residual conventional milk. It also must be processed at the beginning of the day's run. Ingredients must be approved by the certification agency prior to use; non-organic ingredients cannot constitute more than 5% of the recipe, and if any ingredients are available on an organic basis, then the organic ingredient must be used.

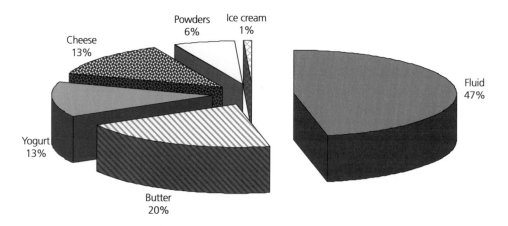

Fig. 8.3   Organic dairy by category.

8.5.1 *Fluid milk processing*

Fluid milk is dairy product in its purest form and constitutes the majority of organic milk consumed today. In some states, you can buy locally produced organic milk directly from the farmer. However, most consumers purchase their milk from a supermarket, which is part of a chain of stores. Supermarkets have the option of carrying glass bottles, cartons of HTST (milk treated by standard pasteurisation at 161.5°F for 15 seconds), cartons of UHT, or plastic jug organic milks. Many carry more than one brand. HTST constitutes approximately 50% of the organic carton milk produced today. This milk has a shelf life of 14–21 days depending on the individual production plant. Ultra pasteurisation (280°F for 2 seconds) is known as UHT. This milk has a shelf life of 45–70 days. Concerns about the degradation of vitamins in the older milk cause most organic production facilities to limit shelf life to 45 days.

Vitamin addition is required in the low-fat milks, therefore all reduced fat, low-fat and skim varieties have Vitamin A palmitate as an added ingredient. In the case of whole milks it is optional for the dairy to add vitamins. The consumer usually has the option to purchase the milk he or she prefers.

Homogenization is the process of applying pressure to milk, so that the fat globules are of a standard size. This causes them to remain evenly distributed in the milk rather than rising to the top. There are homogenised milks (Organic Valley, Horizon and Organic Cow) on the market as well as 'cream tops' (Strauss Family Farms). Non-homogenised/cream top milks are generally marketed in glass bottles.

8.5.2 *Butter processing*

The churning of sweet cream produces butter. In the United States there are continuous churns, which apply pressure and agitation to cream on a continuous stream, extracting the buttermilk and adding salt throughout the process. There also is the older, batch churn method of butter production. The consumer generally does not understand the difference, and both types of product sell side by side on the shelf.

Batch churning is one of the steps required for the production of cultured butter. The slow ripening of cream inoculated with a culture over 16 hours produces cultured butter. The resulting cultured cream is then churned into butter that has a unique flavour.

8.5.3 *Cheese production*

Because of the size of organic dairy, hard cheese production is primarily accomplished through small family-owned cheese plants, which have the size and capacity to segregate the milk and produce a fine cheese. Many cheeses are produced organically: mild and sharp Cheddar, low-fat cheeses, farmhouse cheeses, Colby, Monterey Jack, Pepper Jack, Feta, Mozzarella, Provolone, Parmesan and Romano, to name a few.

Cheese is manufactured by the heat treatment or pasteurisation of milk, which is then cultured, and renneted (for coagulation). Primarily, vegetarian rennets (like *Mucor Mihei*) are used, as genetically engineered rennet is prohibited and vegetarians spurn calf rennet. After the curd is developed and cooked, the whey is drained and the curd is salted, aged and formed. Of course this varies dramatically between the different cheeses. Some cheeses are formed into blocks, and salt is added by placing the formed cheese in brine, other cheeses such as cheddars are matted and cheddared prior to salting.

Salt is an important ingredient in cheese making. It acts as an inhibitor to the culture, slowing acid production, thereby slowing the too rapid development of flavours. This in combination with cooling has a critical effect on the flavour and profile of the cheese. The majority of salt found in the marketplace contains a flowing agent such as 'yellow prussiate of soda', which is prohibited in organic production, and only pure salt can be used. When brines are used, they must be dedicated to organic production only. Organic cheeses generally are aged in a cryovac or plastic bag, to prohibit the growth of mould. Many conventional cheese manufacturers apply mould inhibitors to their cheeses, which is prohibited in organic production.

Organic cheese production was one of the foundation stones of organic dairy just 10 years ago. Today organic cheese accounts for 13% of the milk used for organic dairy products.

8.5.4 *Yogurt*

There are many profiles and styles of organic yogurt today. With close to a dozen brands in the market, it is one of the fastest growing segments in the organic dairy industry, having increased from less than 1% of organic milk usage in 1993 to 13% in 1999.

Yogurt is produced by the culturing of organic milk and solids (such as non-fat dry milk). It is a simple process. Flavoured yogurts present the challenges. The majority of the American marketplace wants a sweet smooth yogurt, which requires sweeteners and flavours. These sources must be organic, and this can have a large impact on the price reflected in the marketplace. Common sweeteners include organic sugar, organic grape juice, organic honey, and organic maple syrups. The syrups create their own issues by adding water to the end product and create a yogurt that is less sweet. Fruits must be certified organic, and flavor enhancers are strictly limited in organic production.

8.5.5 *Ice cream*

Organic ice cream remains a small proportion of the dairy market. With many upmarket ice creams available for consumers, organic ice cream must be of a premium quality. Ice cream in its basic form is milk, cream, sugar and flavour. Premium ice creams contain many novelties, from ribbons of caramel to chunks of fruit, and conventional ice cream manufacturers use many emulsifiers (which are prohibited in organic production) to create a smooth texture.

Sweeteners and fruits, as in yogurt production, are the limiting factors in creating a premium organic ice cream. There are two national companies making organic ice cream at the present time, using less than 0.5% of the organic farm milk produced in 1999.

8.5.6 *Powdered milk*

The ingredient business is important in the organic industry, as it provides milk, butters, cheeses and dried ingredients for the production of other organic products such as ready-to-eat entrées, cookies, confectionery, etc., as well as for use in further dairy production such as yogurt or ice cream.

The drying of milk is an involved process of condensing or ultra-filtration, and spray drying. After drying the powdered milk is then bagged in 25 or 50-lb lined bags. The shelf life of dried milk is up to 12 months. Generally powdered milk is not agglomerated (the process of injecting steam to create an 'instantised' product – easily soluble in hot or cold water). The critical issue with organic drying is isolation, which is achieved by being the first run of the day.

8.6　Conclusion

The organic dairy sector has modelled itself on the conventional dairy world in that it is represented in every feature of the dairy supermarket. From yogurt to milk, from cottage cheese to cheddar, from butter to powdered milk, there is an organic counterpart to each conventional product. The farms range in size from 9 cows to 1000, from family owned and operated to corporate enterprises.

However, with all these similarities, the difference in organic is the treatment of the soil, the air, the water, the animals and ultimately the value of the food produced for the people. In America consumers are becoming more conscious of their need to demand food which supports rural America and is healthy for the farm worker, the store handler and the ultimately the family which it nourishes.

9 Baked Goods and Cereal Products

Andrew Whitley

9.1 Cereal products

Production of organic cereals in the UK in 1998 was an estimated 31 900 t with a farm-gate value of £5.77m (Soil Association 1999, p.10). Wheat was the biggest single UK organic cereal, with 18 000 t divided equally between milling and animal feed (see Table 9.1). Oats and barley followed with 8 100 and 3 800 t respectively, and a small quantity of triticale and rye were grown. Rapid growth is likely in this as in other sectors of the organic market. Key issues for producers and processors include market access, varietal choice and continuity of supply.

9.1.1 Cereal usage

Bread production has traditionally been the main outlet for organic wheat. Oats for meal or flakes, barley for brewing and rye for bread or crispbread are the other main cereals used for human consumption. Spelt, an ancient relative of wheat, is growing in popularity with increasing numbers of allergy sufferers. Several organic cereals – mainly wheat, rye, barley, oats and maize – are flaked for direct consumption in muesli and similar breakfast products or for further processing into cereal bars and flapjacks. Imports of exotics such as quinoa, kamut and amaranth are very small and mainly of interest in special diet formulations, as are non-cereal flours such as chestnut, lupin, manioc and gram.

The inclusion of livestock standards within the EU Organic Regulation (European Community Council Regulation (EEC) No. 2092/91) from August 2000 and the progressive

Table 9.1 Estimated tonnage and farm-gate value of organic cereal crops harvested in the UK in 1998 (source: *The Organic Food and Farming Report 1999*, Soil Association).

Crop	Production (tonnes)	Farm-gate value (£m)
Milling wheat	9 000	1.80
Feed wheat	9 000	1.70
Milling oats	1 600	0.26
Feed oats	6 500	0.98
Processing barley	800	0.14
Feed barley	3 000	0.55
Triticale	1 400	0.24
Rye	600	0.10
Total	31 900	5.77

reduction of the permitted non-organic component in feed rations will fuel demand for all organic cereals suitable for animal nutrition.

Other uses of organic cereals include syrup production (wheat, maize, barley and rice) and brewing. The number of organic beers on the market is growing fast and distilled grain spirits such as vodka and gin have recently made an appearance (see Chapter 11).

9.1.2 *The market*

Historically farmers growing organic wheat have enjoyed a substantial premium for their product. At times, this has resulted in the price of UK organic flour approaching twice that of non-organic. Premiums have been driven more by supply shortages than quality considerations. Recent launches by major bakeries of organic bread products aimed at a mainstream market have accentuated the problem of matching supply with demand. To a farmer considering conversion, this is both an opportunity and a threat.

UK farmers, who in typical years produce wheat of, at best, average bread-making quality, have usually found mills eager for a local component in their grist. Failing that, there has been little problem disposing of poorer quality wheat for organic livestock feed. However, this situation may not continue indefinitely. New supplies of organic cereals in Australia, Eastern Europe and countries of the former Soviet Union are boosting traditional sources in North America and Western Europe. Some of these areas produce high-protein wheats at low cost. One leading English organic miller reported in 1999 that he could buy Australian organic wheat more cheaply than wheat grown in his neighbouring village, even allowing for import levies.

Conversion

When considering whether to convert to organic cereal production, farmers need to bear in mind the influence of these global factors on a market which, though growing, is still very small. Demand for organic cereals is ultimately driven by the organic food market. Any slowing or reversal of growth in this market could result in an oversupply which might weigh heaviest on UK producers of lower protein wheat. A conversion strategy must, therefore, take into account the long-term nature of any commitment to organics, set against considerable uncertainty in the medium-term outlook for prices.

It is clearly not easy for a farmer, once having converted to organic production, to switch quickly back to a conventional system. It is also doubtful whether complete reliance on one type of crop is a sensible way of managing risk. Furthermore, an essential feature of organic agriculture is the control of risk through diversity. Continuous mono-cropping is not sustainable without increasing applications of synthetic fertilisers and biocides which are not allowed in organic agriculture. Systems must be adjusted in a way which effectively reclaims for the farmer the expertise abdicated to the chemist. Crop rotation builds fertility and counters the threat of pests. Involved husbandry replaces routine medication and the philosophy of total control. A necessary extensification may result in an initial lowering of yields, but increasing biological diversity brings greater long-term protection.

Successful organic farmers adopt a similar approach to the food market itself. Cereals form one element of a varied cropping programme whose diversity brings both biological and commercial benefits.

Selling the crop

While selling cereals on the spot market could in the past be relied upon to yield higher prices, there is much to be said for forging links with mills who specialise in organics. As the volume market is increasingly dominated by low-cost producers from favoured parts of the world, the UK organic cereal producer would do well to develop a more specialised niche. If, in the past, the market was hungry only for high-protein bread flours, opportunities are emerging for cereals suited to such applications as pizza, crackers and sweet biscuits. The demand for non-wheat flours also looks set to grow rapidly. The market for home-baking flour is generally in long-term decline, the victim, in the UK at least, of modern lifestyles. However, sales of organic bagged flour are on the increase, suggesting that the dedicated home baker is serious about flour quality. When aiming at a niche, it is advisable to have a clear idea of where the crop is destined to be sold before any seed is sown. On the other hand, the growth of internet trading may provide a convenient way of selling a crop outside the traditional merchant network.

UK market leaders

The leading UK organic millers are Shipton Mill and Doves Farm Foods, both of whom supply significant volumes to the baking trade as well as a range of specialist flours for retail sale (Fig. 9.1). W. Jordan & Co. are market leaders in flaked cereals and bars. With a long-standing commitment to conservation grade cereals, the company has recently launched organic versions of some of its key products.

Some medium-sized mills have converted part of their output to organic in recent years. Of late even the very biggest concerns like Allied Mills and Archer Daniels Midland have entered the field, doubtless with predictable aspirations for market share. A pioneering

Fig. 9.1 Flours from Shipton Mill, one of the UK's leading organic millers.

example of an organic cereal producer who may avoid some of the threats posed by globalisation is Rushall Farm in Wiltshire, which bakes its own cereals into products sold at its busy farm shop.

9.1.3 *Production and processing issues*

As has been suggested above, certain quality issues are of major significance to organic cereal growers, notably choice of variety of cereal and post-harvest treatment. Most modern wheat varieties have been bred for short straw length in order to minimise lodging. These short-strawed varieties provide less shade for adjacent weeds – not a problem when herbicides are deployed, but of considerable significance in an organic system.

A small number of winter wheats (Hereward, Malacca, Spark) dominate UK cereal growing. While well adapted to chemical agriculture and the needs of the dominant Chorleywood bread-making process, they may be less suitable than certain older varieties (Maris Widgeon, Flanders, Avalon) for the more traditional baking methods used by customers of the specialist organic mills. There may be an opportunity for the plant breeders to develop varieties specifically suited to organic conditions.

With cereals destined for human consumption, treatment at and after harvest is crucial. Under organic regulations, careful drying and scrupulous cleanliness are essential because fungicides or pesticides cannot be used. It goes without saying that the general standards of separation and traceability which are required in all organic production and processing must be built in to cereal enterprises from the ground up.

9.1.4 *Breakfast cereals*

Breakfast cereals fall outside the definition of 'baked goods', which will be covered in the next section. Breakfast cereals in the form of simple oat flakes or muesli were among the first organic products widely available, even before the promulgation of EU Regulation 2091/92. In the early 1990s, Doves Farm introduced the first organic cornflakes, followed by Whole Earth and Silbury Marketing with a variety of processed cereal flakes and novelties.

The marketplace at the end of the century was considerably more populated, with ranges from Doves Farm, Whole Earth, Nature's Path, Jordans (see Fig. 9.2) and Alara. Cereals for babies came from Familia, Hipp and Baby Organix. Doves Farm, with its 'Noughts and Crosses', is currently the only manufacturer with a flaked cereal range specifically aimed at children. Oats (especially from market leaders Mornflake), muesli and 'crunchy' cereal, fruit and nut mixtures are widely available, with private label options now appearing.

The major manufacturers such as Kellogg have, to date, not entered the organic market. The flaked cereal process is highly automated and requires significant volumes to be cost-effective. Whereas ingredient availability may not be a problem for the smaller specialist manufacturers, it could be a considerable headache for bigger players for some years to come. That said, there are few technical limitations to producing organic variants of many of the most popular breakfast cereals.

Fig. 9.2 Organic conversion: an organic version of a well-established breakfast cereal from Jordans.

9.2 Baked goods

9.2.1 *Market overview*

Although detailed statistics are not available, the main product areas in organic baking in the UK, by volume, are in-store bakery bread, mainstream sliced bread, speciality breads, and cakes, biscuits and other products.

In-store bakery bread

Organic bread sold through in-store bakeries within supermarkets is rarely if ever produced on the premises. When the EU Regulation became law in 1991 the multiples opted not to seek certification for scratch baking in their in-store bakeries. Although justified at the time on grounds of cost – £ 300 to license a production unit – the more likely reason (apart from the relatively small volume of organic sales) was the complexity of achieving proper separation between organic and non-organic baking in a fast-moving environment. As a result, much of the organic bread sold as 'fresh' over the in-store bakery counter is in fact bought in frozen or part-baked and finished off at the point of sale. Controversy rumbles on as to whether such bread can truthfully be described as 'fresh', given that it may have seen the inside of an oven more than once. Be that as it may, there is still a need to avoid confusion or cross-contamination with non-organic products being made in the same bakery. One ingenious solution adopted by a major supplier was to bake in tins embossed with the word 'organic'; whatever its subsequent treatment after delivery to the in-store bakery, this bread's organic credentials were thus assured.

As long as organic and non-organic production is permitted in the same facility (with due attention to separation and traceability, as enjoined by organic standards), it is open to in-store bakeries to widen the variety of their organic offering by certifying individual premises for scratch production. After two decades in which many craft skills have been lost to the industry, this is unlikely to happen.

Sliced breads

A major feature of the growth of organic baking in the late 1990s was the development of 'mainstream' ranges of white, wholemeal and malted grain breads, presented in the same 800g sliced format as their non-organic counterparts. Several multiples launched own-label ranges, and offerings under the Hovis and Harvestime brands generated significant volume. By early 2000, Harvestime claimed to have more than 70% of the UK organic bread market. If in the early days organic bread clearly demonstrated its wholefood origins, these new products have made a determined bid for the emerging mainstream organic market in terms of price and quality.

Speciality breads

Speciality breads burst on to the UK market in the early 1990s, led by ciabatta, the authentic-sounding slipper-shaped bread invented by a canny Italian some years earlier. In its wake came other Mediterranean breads, rye breads, sourdoughs and breads from all parts of the globe. Suddenly the supermarkets could do something with bread other than discount it below the cost of production. Smaller bakeries like the Village Bakery in Melmerby, Cumbria (see Fig. 9.3), dedicated for years to organic production, were invited to supply major multiples. Some innovative organic breads thus reached a public eager for something more than the handful of traditional organic wholemeals on offer at the time.

Since then, the speciality bread market has matured and segmented. Ciabatta became almost a commodity and can now be found in a variety of baked and 'ready-to-bake' formats, including some without a hint of olive oil in them! Ethnic flatbreads (pitta, naan, lavash) carved out a niche, often using modified atmosphere packaging to offer extended shelf life. The various 'breads of the world' have had a chequered history, those inspired by some genuine knowledge of a baking tradition faring better than some which seemed to amount to little more than the same dull dough with different bits in.

In terms of volume, organic speciality breads are always likely to remain a niche, given the predilection of the British public for less challenging fare. They do, however, provide

Fig. 9.3 Organic pioneers: the Village Bakery, Melmerby, has succeeded in bringing artisan breads to major multiples.

a useful testing ground for ideas which may later converge with the mainstream; and they are also one of the few realistic entry points for the smaller organic baker eager to dip a toe in the fast-running waters of multiple grocery.

Cakes, biscuits and other products

Organic cakes have consistently played second fiddle to breads, possibly because until recently no major manufacturers were licensed for organic production. Some smaller concerns (Doves Farm [see Fig. 9.4], Sunnyvale, Whole Earth, the Village Bakery) have supplied the independent trade for many years; a few of their confectionery products have also made it into the multiples.

After one or two unsuccessful launches in the early 1990s which were dogged by the high price and patchy availability of raw materials, organic pizza now seems likely to establish a place within what is a mature and well-defined sector. Most sales of organic pizza are likely to be of frozen or chilled prepared product. Ensuring separation within a take-away parlour also making up non-organic pizza might be considered risky. But given the dynamic growth of all things organic, it cannot be long before dedicated organic pizza (and other fast food) outlets make their mark. Already in the UK the Pizza Piazza chain with 50 pizzerias has announced its intention to convert to a fully organic format under the name Pizza Organic.

Fig. 9.4 A successful organic alternative from Doves Farm.

Organic biscuits have been relatively slow to catch on in the UK. Suitable ingredients have been hard to source until quite recently and predicted sales volumes have failed to entice bigger players in what is a highly automated sector. This has left some room for continental offerings. A successful biscuit brand, Duchy Originals (see Fig. 9.5), based on cereals grown on the Prince of Wales' organic farm, has recently achieved full organic status.

Cereal bars and flapjacks are an area of baking which has seen considerable growth. Organic versions are offered by some smaller bakers. Other categories such as bagels, muffins, croissants and Christmas puddings all have organic offerings. None has yet achieved anything like the penetration of organic bread. Some products, for instance decorated celebration cakes, are rarely if ever seen in organic form, either because an

Fig. 9.5 Duchy Originals biscuit range, with a pronounced gourmet appeal, are now fully organic.

ingredient (for example, icing sugar) is hard to make organically, or because the process traditionally requires additives which are not permitted, or because there is little demand for an organic variant of such a specialised item (see Fig. 9.6).

9.2.2 *Operational issues*

Separation

The first issue for a baker contemplating organic production is how to guarantee separation between organic and non-organic activities. Although the EU Regulation at present requires dedicated facilities only for fruit and vegetable packing lines, complete physical separation is recommended for all organic processing and may conceivably be required at some future date. In the meantime, production may take place in mixed facilities, providing that adequate systems are established to achieve separation. A separate organic production area

Fig. 9.6 An award-winning confectionery product from the Village Bakery, Melmerby, demonstrating a successful collaboration with another innovative brand, Green & Black's chocolate.

can be created within a larger plant, or organic batches can be run on machinery which is otherwise used for non-organic processing. Bleed runs must be used to purge continuous process machinery prior to an organic run. Batch processing of organic ingredients must take place after a thorough cleandown of all machines and utensils.

Traceability

Traceability must be built in to organic processing in order to achieve full accountability for all organic ingredients purchased and all organic products sold. Quality assurance systems such as ISO 9000 embody appropriate disciplines. All ingredients for organic products must be accompanied by proof of their organic status. This can take the form of a general warranty from a supplier who is certified as an organic packer or processor, or specific documentation provided by a supplier who is not so registered.

Training

Any system is only as good as the people operating it. Training is therefore vital. In small and medium sized operations especially, it is likely that ingredients (for example, flour, oil or fruit) of identical appearance will be used in organic and non-organic formulations. The inspector from the organic certification body will seek evidence of carefully segregated stores and a general awareness of the importance of separation among the workforce.

Authenticity checking

Reference has been made elsewhere (Chapter 2) to the rules governing multi-ingredient products with either 70–95% or over 95% organic content. The aspiring organic baker will spend many happy hours poring over the minutiae of permitted agricultural ingredients of non-organic origin and lists of additives and processing aids.

Particular note should be taken of the current EU requirements (as embodied in the various national standards) which prohibit the use of genetically modified organisms (GMOs) within the organic supply chain. This is not simply a question of eliminating GM ingredients themselves; the regulations also prohibit materials in whose production GMOs may have been involved. This rules out the use, for instance, of GM enzymes, even if none remain in the final product.

Organically certified ingredients are, *ipso facto*, GM-free. But the same cannot be said for additives or processing aids which come from conventional sources. The obligation is on the organic baker to gain assurances from suppliers that GMOs have not been used in the manufacture of all such ingredients.

9.2.3 *Technical limitations*

There are various technical issues which affect organic production and which may limit the scope of product development.

Supply chain

In a rapidly expanding sector, there is bound to be some mismatch between demand and raw material supply. Given the stringent demands of organic certification and the fact that it takes at least two years for land to come into organic production, it is hardly surprising that shortages (and gluts) occur. Even if supplies of a particular ingredient are available, scarcity may push the price to unacceptable levels. Pioneering organic processors have long battled with high raw material prices. Until recently in the UK, organic raw cane sugar cost almost three times as much as its conventional equivalent and it is still not uncommon to pay 50–100% more for organic ingredients.

This is partly a reflection of higher costs of production; but the main explanation must lie in the historically small volumes traded. As new farms come on stream, this factor should diminish in importance. Nevertheless, a processor seeking to develop a product which will use any significant volume would do well to secure supplies well in advance. Even when such precautions are taken, however, disaster can still strike. Some of the more unusual ingredients may be produced by only one or two farmers, with potential consequences for continuity of supply. For example, a crop failure in 1999 led to a complete cessation of the supply of organic lupin flour, important in the production of gluten-free bread.

In extreme circumstances, a processor can apply to UKROFS for permission to use a non-organic substitute, even if the ingredient does not appear on the permitted list of agricultural ingredients of non-organic origin. Proof must be supplied in the form of a statement from suppliers that no organic material is available. Label declarations must always reflect the true status of each ingredient. Derogations to use non-organic equivalents are normally issued with a duration of 3 months, after which a further application must be made. Such derogations are discussed at regular EU meetings, giving all countries a chance to identify previously unknown sources of organic supply.

An interesting controversy arose in the early 1990s when an English baker applied for permission to use non-organic Greek olives in an organic Greek olive bread. At that time, there were no certified organic olive products coming out of Greece. The Spanish delegation in Brussels argued that there were plenty of organic olives in Spain. But the baker was able to continue using non-organic Greek olives on the grounds that it might be misleading to use Spanish olives in a product claiming Greek provenance. Happily, the development of the organic market throughout Europe has put an end to such disputes.

As supplies improve, the number of agricultural ingredients which may be used from non-organic sources (within the 5% tolerance permitted under the EU Regulation) is bound to reduce, perhaps eventually to zero. It would be rash to base any medium-term strategy on the continuance of the current list.

Secondary processing

Even now, however, it may not be possible to find organic ingredients in the precise format required. Organic ground almonds may be available, for instance, but not blanched split almonds. It may be necessary to make special arrangements for further processing of an organic ingredient available in a basic form. This will involve the search for a processing plant which itself has organic certification. There may be cost implications in secondary processing of relatively small volumes.

Additives and processing aids

Such difficulties are certainly diminishing as the whole organic supply chain matures. But product developers will still come up against the limited number of additives and processing aids permitted in organic production. Although special cases have been and continue to be made for the use of specific aids in order to permit production of certain familiar ingredients or products, the whole thrust of the EU Regulation has been to respect the aspirations of consumers who want organic food to be of significantly higher nutritional quality than conventional fare. Hydrogenated fats, for example, are not permitted under organic standards because of evidence that their consumption may harm health. Likewise, organic bread improvers are limited to ascorbic acid, lecithin and certain (non GM) enzymes.

There is pressure from larger manufacturers to extend the permitted list of processing aids and additives to include whatever is required to enable organic production of virtually every type of product. A foretaste of possible future conflicts was seen when the US Department of Agriculture attempted to allow GMOs (as well as other elements banned in Europe) into the draft national organic standards in America. An unprecedented outcry from the organic trade and ordinary consumers forced the USDA to backtrack. At stake is the question, vital to many in the organic movement, of whether organic is just another processing option or a food production system with wider environmental and social implications.

9.2.4 *Market issues*

In general, if the early days of organic baking were pioneered by companies with a clear commitment to food quality in the full nutritional sense, recent developments have been all about colonising the 'mainstream'. While it is understandable that major retailers should seek to produce organic versions of their best-selling lines, there are those who question the wisdom of such a strategy. In the first place, as has been discussed, organic processing standards may place restrictions on additives or aids with the result that the organic variant differs slightly from its non-organic precursor. This might cause disappointment in a consumer expecting exactly the same eating experience. Furthermore, it is not immediately obvious how the wider significance of the organic project (with its commitment to health and wholeness) is reflected in products which ape some of the sugar, fat, and additive-laden effusions of the global food industry. Cut adrift from nutritional integrity, organic becomes just another brand. The food-aware consumers who have driven organic sales and who have shown that they are prepared to pay more for the organic guarantee, may not in the long term take kindly to companies who make no room for any ethical baggage on their commercial bandwagon.

The market power of the British multiple retailers is undisputed. Having decided to respond to public demand for organic products, they now account for some 74% of organic sales (Soil Association 1999, p. 27). Smaller organic producers, once courted by multiples concerned to attract the organic customer (whose trolley is worth significantly more than average), now find themselves competing with large manufacturers who can offer an organic alternative within a category in which they may have a dominant position.

British supermarkets have hurried to apply their own brand names to many organic products, perhaps unaware of a potential contradiction: if the company's own brand now embraces organic production, with all its positive messages about how food is produced with respect for natural systems and without chemicals, how can it credibly also be applied to non-organic products emanating from a different – and contradictory – food system?

One response to the growth of private-label and other competitive activity must be for small and medium sized organic producers to seek out the new opportunities which are emerging in the form of specialist organic stores and farm shops, organic local markets, mail order via the internet and so on.

9.2.5 *Distribution and packaging*

Some of these emerging markets may offer an alternative to the ever-rising costs of conventional distribution. Cereal products and baked goods are, generally speaking, bulky and heavy in relation to their value. Breads and morning goods require daily distribution. To cap it all, the organic market is perhaps more than usually sensitive to the issue of 'food miles', the overall distance which food products travel from farm to processor to consumer. In reality, the 'ecological footprint' of a fully-laden articulated wagon may be significantly less than that of a baker's van making multiple drops in a rural area. The significant fact is that this issue is on the agenda, for the organic processor perhaps more than his conventional counterpart.

For similar reasons, packaging of organic products must be designed to be consistent with food safety while avoiding unnecessary waste of resources. It is unclear whether something akin to the German 'green point' system will be implemented in the UK, whereby producers of packaging have to take responsibility for its reuse or recycling. Suffice it to say that the organic product developer should seek to use packaging materials and methods that are consistent with the wider principles of the organic food system.

9.2.6 *Future trends*

What of the future? The organic cereals and baked goods market seems likely to continue its recent rapid growth for some time to come. In parallel with the move to produce more 'mainstream' products, other trends may emerge.

Fair trade

Organisations like Oxfam, Traidcraft, Equal Exchange and Café Direct are increasingly involved in the food market, importing ingredients or finished products from Third World producer groups who are paid a fair price, usually guaranteed to be significantly above world market rates. Coffee, tea and chocolate have led the way, some brands certified by the Fair Trade Foundation. More baked products – cakes, puddings and cereal bars – are appearing, designed to use fairly-traded fruit, nuts and sugar. There is not, at present, a complete convergence between fair trade and organic. However, those working in the field are conscious that for most concerned consumers, the idea that a product should be produced with concern for *either* the environment *or* the growers but not both is contradictory.

Special diets

There has been a significant increase in the incidence of people reporting allergies, principally to wheat, but also to other gluten-containing cereals (rye, barley and oats). While cynics may see a correlation between the prevalence of allergies and the growth in the number of allergy clinics, there is no doubting the effect on the potential market for cereal products. There is a view that allergies to wheat may be the result of changes in its biochemical structure due to intensive hybridisation over recent decades. If this is so, it is a further pointer, if such were needed, to the dangers of genetic modification. But it also suggests a possibly fruitful line of research into more traditional grains which may have preserved important nutritional features. Specialist organic farmers, millers and bakers might learn from the growing popularity of spelt (*Triticum spelta*) to develop new products.

Dedicated processing

Just as the consumer would be surprised to learn that 20% or 30% of the feed ration of organic cattle or poultry can be non-organic, so the realisation that organic multi-ingredient products can legally be produced in non-dedicated plant might easily raise more than a few journalistic eyebrows. While clearly there are considerable obstacles to full separation, it would be advisable for anyone involved in organic production to reckon with the possibility that, one day, it may be mandatory. If this seems far-fetched, it is surely no more so than the prediction that nearly a third of European agriculture could be organic by 2010 (Soil Association 1999, p. 2).

References

European Community Council Regulation (EEC) No 2092/91.
Soil Association (1999) *The Organic Food & Farming Report 1999*. Soil Association, Bristol.

10 Other Organic Processed Food

Andrew Jedwell

10.1 Introduction

The object of this chapter is to outline the factors influencing the intending producer of processed organic foods and drinks not covered elsewhere in this book. The term 'processing' of raw materials of agricultural origin encompasses a range of activities from simple operations, such as trimming and grading vegetables, or simply breaking bulk, through to complex factory processing operations, restaurants, catering etc. It is to the factory-based processor that this chapter is primarily addressed.

10.2 Overview

The market for processed organic foods has undergone a dramatic change over the recent past and is still evolving at a rate which shows no signs of slackening. The effect of this has been a huge explosion in the total value of organic products sold, to the point where J. Sainsbury cited organic as the fastest growing area of sales in 1999. Estimates of the overall rate of growth vary, but the 1999 Soil Association report indicates a compound growth of over 40% for the four years 1996–2000. This growth has been fuelled by a large rise in the total number of organic lines stocked, particularly in processed products, many of them with complex multi-ingredient recipes. This broad shift reflects the cycle of adoption by organic consumers, who typically start with primary produce, such as vegetables and milk, and then work their way, via staples such as dairy, meat and bread, up the ladder of value-added products. Products such as ready meals, convenience foods, confectionery and alcoholic beverages represent the top rungs of the ladder.

The total number of organic lines stocked by the leading grocery multiples was approximately 300 at the end of 1998 and this had doubled to approximately 600 by the end of 1999. Some estimates suggest that by 2001 the number of stock lines for a multiple retailer that is committed to organic will exceed 1000. It is already the case that the largest independent health food retailers stock 3000 certified organic lines. Since the majority of primary staples have already been on sale for some time, the largest part of this growth in number has been, and will be, supplied by processed products.

The route to market for these products falls broadly into four main categories:

(1) Dedicated organic brands that previously were of interest only to specialist outlets, such as health food shops, have found wide-scale multiple distribution. Examples are Meridian Foods, Baby Organix, Whole Earth and Doves Farm (Fig. 10.1).

(2) Well-known mainstream grocery brands have launched their own organic sub-ranges: examples are Baxter's soups and Libby juices.

(3) A few companies, perhaps concerned about possible conflict of values with their existing brands, have launched separate organic brands. An example of this is Seeds of Change, which is under the ownership of the Mars family, who also control Uncle Ben's and Dolmio, among other brands.

(4) Own-label. This has perhaps been the fastest growing area, following the first rush of branded organic products onto the shelves, and already by the end of 1999 approximately 50% of the range of J. Sainsbury's organic products was under own label.

Organic products are already present in virtually all sections of the market and, although most surveys consider that they have so far achieved less than 4% of all food sales, there are examples of much higher degrees of penetration. In some of the multiples, it is estimated that 10% of all fresh fruit and vegetables sold are organic and this market share has already been exceeded for organic prepared baby foods: for this latter sector there are some multiple grocers where approaching half of all prepared baby food sales are organic. The chilled ready meals sector, one of the most dynamic categories of conventional product in recent years, is starting to see a significant impact from organic products and there is no reason to doubt that retailer own-label will assume its customary dominant position in this sector.

The extent to which organic options are becoming available in most sectors is shown by the increasing presence of organics in such diverse areas as ice cream, prepared salads, a wide choice of wines and even organic spirits, mixers and cola. From the sybaritic pleasures of chocolate truffles, to organic food for the family pets, there is no doubt that a complete and rounded organic diet is now possible (see Figs 10.1–10.8). There are,

Fig. 10.1   A selection of organic products from Doves Farm.

Fig. 10.2 Natural Selection organic mixer drinks.

however, areas where coverage is still patchy, perhaps indicating that problems on the supply side are still not yet fully resolved. For instance, the availability of organic herbs and spices is still limited, pointing to problems for both herb packers and also manufacturers who need a consistent and reliable supply in order to meet the demands of their multi-ingredient products.

This dynamic growth has propelled the number of registered processors in the UK to well over 1000, ranging from the simple packer through to processors of complex products such as restaurant meals and ready-to-eat chilled or frozen dishes. A large proportion of these manufacturers also have conventional (non-organic) operations and have entered organic as a new development, encouraged by the overall growth of the market or often at the behest of their supermarket customers. While there is nothing to prevent this dual-mode operation, so long as the regulations are scrupulously observed, there is also a steady trickle of entrants to the market who are running organic-only factories or plants, and this trend seems likely to gather momentum.

The requirements of organic processing are additional to those of existing legislation and good manufacturing practice. None of them should trouble the competent manufacturer and existing systems, such as Hazard Analysis Critical Control Point (HACCP), will be a great help in achieving organic compliance. Nevertheless, any company will have to address a number of issues when entering the organic market. While most of them are easily surmountable, all such decisions will have to be taken so that they conform to the standards of the sector body which has been chosen. The obvious corollary, and one often overlooked, is to ensure that within the company there is someone who is familiar with, and understands, the regulations. Since all product and process developments grow more costly to correct, the further downstream they progress; this front-end investment will repay itself very quickly. It will also indicate that the processor fully accepts the need to work within the organic rules: a reluctance to do so will lead to unnecessary frustration. The

Fig. 10.3 Clipper organic instant coffee.

regulations are designed to support a sustainable system of primary food production, and to define complementary processing inputs and methods, not to ensure that all conventional foodstuffs can be replicated in organic.

10.3 First steps

Having decided to enter the market, the EU Organic Products Regulation (EU 2092/91) requires that anyone who wishes to produce organic food must register with a certification body. This body is responsible for ensuring that anyone who wants to produce organic food understands the legislation and has the necessary procedures and systems in place: the certifying bodies are in turn policed by UKROFS. Local trading standards officers enforce the regulations in the marketplace.

The processor therefore arrives at the point of making a decision as to which sector body with which to certify. This should be an early decision because the standards of any one of the sector bodies may make additional requirements to those which are in the Regulation. A copy of the sector body standards will therefore be needed to start planning both process and products. In addition, the sector body will probably provide a starter pack and also an advisory service for members on how best to ensure compliance.

Fig. 10.4   Clipper organic tea bags.

There are currently eight sector bodies that are accredited by UKROFS, all of which offer certification to processors. Among the factors that may be taken into account when making a choice are:

- *Quality of service*. Does the sector body provide a prompt and comprehensive response?
- *Recommendation*. Has it been right for other companies in the same product area?
- *Recognition*. Does its symbol offer your product sufficient credibility in the marketplace?
- *Customer requirement*. Do your key customers have a particular symbol preference?
- *Special factors*. Do you, for instance, have a geographical requirement (for example, SOPA for Scotland) or a particular affiliation (for example, BDAA for biodynamic processors)?
- *Cost*. Cheapest may not necessarily be the best.

Once the decision has been made, the procedure is likely to follow the following route:

(1) Fill out the initial application form supplied by the certifying body. Note that on the initial application it is necessary to list the recipes of the anticipated launch range of products. It is easy to make changes and add additional products at a later date.
(2) The certifying body sends an inspector to the manufacturing premises. In an operation where both organic and non-organic products are manufactured the major point of concern is that there is no contamination from non-organic to organic. All systems and physical procedures need to be designed to achieve separation by space (production lines dedicated to organic production) or time (organic production following a full cleandown, using cleaning methods and compounds approved by the certifying body). There are a limited number of circumstances, particularly large-scale process plants, where separation may be achieved by purging with organic raw material, which is

Fig. 10.5 Green & Black's organic chocolate bars.

then discarded once it has effectively removed all non-organic residues from the plant. This process is seen, for instance, in some sugar factories. As well as the production lines, storage areas for raw materials and finished product should be considered, in order to ensure that organic materials are clearly identifiable and there is no risk of cross-contamination. It is also likely that, if third party storage and distribution is used, the sector body may wish to include these in the audit. Table 10.1 shows in summary form the considerable depth of information now required by such an audit, in this instance by Soil Association Certification Ltd.

(3) The inspector submits a report to the certification committee of the certifying body. If the report is approved, the certifying body issues a certificate and the operation can then begin supplying organic products and use the certifying body symbol. Note that it is illegal to produce organic food and drink without first going through this procedure. The issue of the certificate may be subject to a written agreement by the licensee to address various minor non-compliances. A general warning about timescale must be issued at this point. The continuing growth of organics has tended to stretch the administrative capacity of the certifying bodies and, at the time of writing, a 12-week lead time was the norm for the period between first application and final certification. It is well worth checking out the timescale when first looking at the whole registration process in order to ensure that the product launch plans are realistic and that there is no risk of trading illegally. Some manufacturers have considered it a worthwhile investment to carry out a self-audit, prior to the first sector body audit, in order to ensure a successful and timely outcome.

(4) The certifying body carries out an annual inspection of premises, systems and production records to ensure that all of these procedures are followed.

Table 10.1 Example of categories used in an inspection report (source: Soil Association Certification Ltd processing and packing inspection report; used with permission).

1 Introduction	**5 Packaging materials**	**10 Pest control**
	• Retail	• Contract
2 General	• Wholesale	• Rodent control
• EHO regulations/other		• Insect control
statutory requirements	**6 Goods in and storage**	• Fumigation
• Standards	• Incoming transport	• Storage
• Training	• Check of organic status	
• Formal OA system	• Storage areas	*Pest control records*
	• Dedicated status	• Manifest
3 Product lines and raw		• Pest control records
materials	*Records*	
	• Incoming documentation	**11 Importing**
New products	• Goods in log	
• Details		*EU countries*
• Marketed before licensed	**7 Production process**	• Details
	• Description	• Certification equivalence
Single ingredient products	• Dedicated status	(livestock standards)
• Comply with standards	• Separation	
	• Bleed runs	*Third countries*
Multi ingredient products	• Processing equipment	• Details
• 95–100% organic	• Storage bins	• Certification equivalence
• 70–95% organic	• Duration of runs	• UKROFS imp. auth.
• Organic ingredients	• Frequency of runs	• EU certificates
• Non-organic ingredients		
• Non-agricultural ingredients	*Production records*	*Other documentation and records*
• Processing aids	• Production runs	• Details
	• Quality control	• Certification equivalence
Imported products	• Traceability	• UKROFS imp. auth.
• Certification equivalence		• EU certificates
	8 Goods out and transport	
Records/documents		**12 Record keeping and**
• Product specification sheets	*Goods out*	**documentation**
• GMO guarantees	• Dedicated storage areas	
• Suppliers' certificates		*Complaints register*
• UKROFS derogations	*Transport*	• Complaints register kept
	• Method	
4 Labelling	• Sealed containers	*Sales records*
• Samples	• Hygiene	• Retail sales
		• Delivery notes
Retail 95–100% organic	*Goods out records*	• Sales invoices
• Reference to organic	• Output records	
• Ingredients panel	• Outgoing documentation	*Accounting records*
• Use of symbol		• Accounting records kept
• EU code	**9 Cleaning procedures and**	• Audit of accounts
	hygiene	
Retail 70–95% organic	• Details	*Stock records*
• Special emphasis	• Materials used	• Raw materials
• Ingredients panel	• Storage	• Finished goods
• % organic ingredients		• Stock reconciliation – sample
• Size/colour	*Cleaning records*	audit
	• Written cleaning schedule	
Wholesale labels	• Records of cleaning	**13 Audit**
• Wholesale labels		• Details of any audits
		• Acceptability of audit
Conversion labels		
• Wording		
• Symbol		

Fig. 10.6 Green & Black's organic chocolate-coated almonds.

The sector bodies will be very thorough in their certification process because it is imperative that the credibility and integrity of their symbols is maintained; this is the reason their attitude may sometimes seem unbending.

The charges involved will vary from sector body to sector body. Inspection visits will normally be charged for separately, and in addition there will be a fee payable to the sector body for the use of their name and/or logo on the pack. It is legal not to show such a symbol but just to detail the sector body's code on the packaging. In these circumstances, the manufacturer will have to consider the possible cost advantages against the likelihood that they would forfeit any advisory services and also the benefits of using the sector body logo.

Fig. 10.7   Juniper Green organic gin.

10.4 Product issues

The Organic Products Regulation EU 2092/91 divides organic multi-ingredient foods and drinks into two categories depending on the proportion of organic ingredients present:

- *Category 1 Organic*. This covers products which contain a minimum of 95% organic ingredients by weight. Products can be labelled 'organic', for example, 'organic cornflakes'.
- *Category 2 Special Emphasis*. This covers products which contain 70–95 % organic ingredients by weight. Product can be labelled 'Made with organic ingredients', for example, 'tomato ketchup made with organic tomatoes'.

EU regulations should be studied carefully because there are a number of other points which have to be incorporated into the labelling. Indeed, the 'organic cornflakes' example quoted above may be held to be inconsistent with the Regulation although it is increasingly the format that is adopted since it is more directly relevant to consumer perceptions than the perhaps more legalistically correct 'cornflakes made from organically produced maize'.

Fig. 10.8 U.K.5 organic vodka.

The Regulation also includes the following criteria:

(1) Annex VI of the Regulation specifies those food additives (Section A), processing aids (Section B) and non-organic ingredients of agricultural origin (Section C) which can be used within the 5% non-organic constituents. Note that these are lists of positive approval and processors cannot use anything that is not specifically identified under this Annex.

(2) If a particular ingredient is not available in organic it is possible to apply for a derogation to use the non-organic version but these applications are being assessed ever more rigorously.

(3) Water and salt are excluded from any percentage calculations and, if products are used in a concentrated or dried form, the reconstituted weight must be taken into account, assuming that reconstitution occurs during the manufacturing process.

(4) The same ingredient cannot be added to a product in both organic and non-organic form.

(5) The use of irradiated or genetically modified ingredients is specifically prohibited in organic food. This will necessitate a rigorous background check of any non-organic ingredients in order to obtain firm undertakings that they are GMO free.

These regulations are likely to lead a conventional food manufacturer to adopt a tailored approach to organic recipe development. As an example, a careful scanning of the Regulation will show that only non-chemically modified starch is allowed, whereas very highly specific chemically modified starches are the norm in conventional processing. This is likely to mean that the development process will have to include examination of non-chemically modified alternatives. Another example is chopped tomatoes, normally used in recipe dishes in the conventional industry, which are treated with calcium in order to retain their texture; this is not an option available to the organic manufacturer.

The issues of cost and availability must also be considered. In general, although not invariably, organic ingredients are more expensive than their conventional counterparts. Although organic ingredients are becoming available in larger quantities, demand is also rising sharply and, while some premiums are falling, some are rising. Availability may often disappear for part of the crop year and this problem is certainly more likely at the tail end of the crop year than with conventional product. This is likely to oblige the purchasing manager to take out annual purchase contracts in order to ensure continuity of supply and consistency of quality. The purchasing department will also have to take into account reliable and appropriate certification and also the organic traceability of materials. It is also likely that the purchasing and stock control departments will be involved in the annual audit, much of which is based around the principle of ensuring that the organic material inputs are consistent in quantity and use with the output and sale of finished organic products. While this may already be part of the existing stock control system, for many smaller companies it may be an additional requirement and one that will have to be met.

The rapid growth of the industry is bringing many suppliers of conventional products into the market and, while many of them are very thorough in their approach, there have been a number of instances recently in which ingredient suppliers have not demonstrated the competence in organic that is required. One of the main issues that manufacturers should keep in the forefront of their consideration is whether the raw materials they purchase originate from outside the EU. Even though materials may have been certified as organic in the country of origin, they will still have to go through an import approval procedure which is managed by the UKROFS secretariat, part of the Ministry of Agriculture, Fisheries and Food (MAFF). It must be confirmed by the supplier that third country imports have been passed through the UKROFS procedure: this has not always been the case. A similar caution applies to manufacturers importing their own raw materials from outside the EU. The UKROFS procedure must be adhered to and, depending on a number of factors, this procedure is likely to take weeks rather than days.

Having developed and launched a product that meets all these requirements, and also one which tastes good, the proud manufacturer might want to consider entering the product for the prestigious Soil Association Organic Food Awards, which are held every year. Categories of potential interest to processors are:

Sausages; Prepared meats; Dairy (including ice cream and yoghurts); Cheese; Flour; Bread; Cakes, pastries, biscuits and puddings; Confectionery and snacks; Breakfast cereals; Condiments, savoury preserves and spreads; Sweet preserves and spreads; Teas, coffees and beverages; Non-alcoholic cold drinks; Alcoholic drinks; Store cupboard staples; Soya foods; Convenience foods; and Baby foods.

The recognition bestowed by such awards is a source of justifiable pride in an excellent product and also confers an accolade which is a considerable advantage in this exciting marketplace.

References

MAFF (The Ministry of Agriculture, Fisheries and Food) (1999) MAFF Consolidated Version of Council Regulation (EEC) No 2092/91. Her Majesty's Stationery Office, London.

Soil Association (1999) *Organic Food and Farming Report 1999*. Soil Association, Bristol.

11 Organic Alcoholic Drinks

Renee Elliott

11.1 Introduction

When Louis XIV said that 'a meal without wine is like a day without sunshine', he meant organic wine. Although many people think of organic as a recent innovation, there was a time, not that long ago, when there was a simpler approach to growing and producing all our food and drink. Organic wine is nothing new either, but follows traditional grape growing and winemaking techniques. This applies to all alcoholic drinks, when the ingredients for wine or beer or spirits were grown in a natural way and turned into beverages with little or no dependence on a chemical armoury. Our grandparents would have mixed copper and sulphur to protect their vines, unlike the winemaking generation of today who can use a range of chemical fertilisers and pesticides.

Organic wines make up the majority of the organic alcoholic drinks market today and therefore take up the main part of this chapter. There is no worldwide standard for organic winemaking, and even though there are restrictions on what may be used in the vineyard and in the winery, the guidelines do not set out what a grower, faced with no or very few chemical crutches, is supposed to do. This chapter sets out to give practical guidelines for those interested in organic viticulture and vinification, to show that there is a more environmentally friendly way forward and to illustrate how big wineries are successfully producing organic wine.

Small family growers who aim for pure, fresh wines with varietal character today produce the majority of organic wines on the market. Some big players are on the scene, too, which banishes any thought of organic wine being unfashionable or hippyish. Multimillion-dollar companies such as Fetzer in California and Penfolds in Australia are committed to organic winemaking, paving the way for an expanding market. Organic wine has made great strides in the last decade. Not only is there more choice, but quality has improved tremendously. Although it is difficult to obtain definitive figures on the number of organic wines available, it is estimated that there may be 8000–10000, with around 500 of those on sale in the UK.

A major tasting of organic wines was held in the UK in the late 1980s, when the panel, consisting of professional and semi-professional tasters and consumers, found the wines on trial at that time disappointing and expensive. Organic wines have improved in quality, perhaps as winemakers are relearning traditional methods. Today, organic wines are taking top awards at the International Wine Challenge in London.

It is important to note that calling wine 'organic' does not guarantee that it is 100% pure and free from pesticides and chemicals. The water the vines will use is likely to carry some pollution, there is the chance of spray drift from conventional vineyards nearby, and some chemicals are allowed in organic winemaking. Principally, organic wine strives to be a cleaner, higher quality product, and stricter controls will develop as the organic wine movement grows.

11.2 Conventional wine

If we take a look first at conventional wine – including champagne and sparkling wines – and the chemical safety net of modern, intensive farming, we can see the cycle in which winemakers can get caught in the vineyard. A vast number of chemicals are permitted in conventional viticulture. The rules concerning how they are used and in what quantity are decided by the appropriate authority governing wine production in each country. To create cheap wines, one needs cheap methods – in the vineyard and in the winery – and intensive use of chemicals is often tied to mass production. Organic or biological agriculture and horticultural systems, however, as described by the Soil Association, 'are designed to produce food of optimum quality and quantity'. There is no such thing as good cheap wine.

Conventional methods in the vineyard can result in vines that are not integrated with a living, vibrant soil. Because of this, they do not have natural resistance and vital strength and lack innate defence against pests and diseases. These dangers become a more serious threat than they would to a strong and healthy plant. Chemical fertilisers weaken the vine so that chemical fungicides and pesticides are necessary to protect it, thus reinforcing the chemically dependent cycle and disturbing eco-systems. So, in a nutshell, the conventional grower can end up with depleted soil and weak plants that are more susceptible to pests and diseases. This happens for a number of reasons.

Grown as a mono-crop, without the benefit of crop rotations, vines strip the soil of the same nutrients every year. Diseases, weeds and pests that affect the vine are allowed to become entrenched in the area. Vines become locked into the cycle of chemical treatments and feeding as the soil continues to be broken down. Because the soil is depleted, the vines become malnourished. Chemical fertilisers are then necessary, but instead of feeding the soil, the chemical fertilisers (which are soluble salts) do not enrich the soil, but directly feed the plant. The vines then live in the earth, but no longer take sustenance from it.

Nitrogen is not 'fixed' in the soil through green manures and nutrients and not ploughed in from compost. As a result, the roots are closer to the surface, waiting for the nitrogen fertiliser to be dumped onto the ground. If too much nitrogen is used, the vine will take up more water than it needs (if water is available), creating in the plant structure fat and juicy cells with thin cell walls. This fatness weakens the plant and is particularly attractive to pests. This compromised plant is then attacked by pests, forcing the grower to go into the vineyard with a chemical pesticide and spray, killing the pests that are now attacking the vine.

But nature is resourceful and adapts to the poisons pumped into the environment. The pest ultimately will develop immunity to the chemical used against it, which creates two problems. Firstly, the temptation is to use the chemicals more frequently to get the same effect. Secondly, another stronger, more effective chemical will ultimately replace the one that no longer works. The process in which pests develop resistance to pesticides is accelerating, which must be worrying for those who are tied into this system. When man competes with nature, it is unlikely that he will win, and likely that he will cause damage in the process.

Fungal diseases present another danger to the weakened plant. There are no cures for fungal diseases, and sprays and/or systemic fungicides are used to try to keep them under control. However, some conventional growers have instituted a programme of 'insurance' spraying whereby vines are sprayed with fungicides every two weeks through the growing season.

The use of weedkillers has also been on the increase over the last 20 to 30 years. Although it is difficult to prove specifically which chemicals may be carcinogenic, there are increased levels of herbicidal residues in drinking water and concerns over many of the synthetic substances being used. Every year, governments in various countries withdraw from use chemicals that have previously been used on our food supply when proof of their toxicity is established. The classic example is DDT, which was used as an insecticide from 1939. Over a period of 50 years almost five billion pounds of it have been used across the world. It became prohibited in the US in the 1970s and in the UK in 1984 because of links to deaths, cancer, allergies, infertility, problems in fetal development and major diseases of the immune system.

Out of the vineyard and into the winery, we discover a raft of additives permitted under EU law. Grapes that have been grown in a chemical environment will have little surviving natural yeast, as fungicides will have destroyed most of them, so to start the fermentation, cultured yeasts are often introduced. Then the winemaker uses a range of chemicals to correct problems that have developed in the vineyard and resulted in poor quality grapes. There are permitted chemicals for acidifying, deacidifying, clarifying, stabilising, preserving, antioxidising and so on. The worse the grapes, the more intervention is necessary to try to produce a passable wine.

Sulphur dioxide (SO_2), the additive most commonly used and thought of in winemaking, can be used when the must is starting to ferment, to kill the yeasts, to fix the wine, to stop the fermentation, before bottling and to sterilise equipment. SO_2 can cause headaches, stomach-aches, hangovers and acute reactions in asthma sufferers.

11.3 Organic regulations

Most organic organisations in different countries belong to the International Federation of Organic Agriculture Movements (IFOAM) and follow its basic principles, but there is no single inspection scheme for all organic wines. Because of differences in technology and science, combined with differing climates, the details of a country's specification will vary. Indeed, within one organisation, regulations may vary from region to region as well as reflecting various styles of wine.

To put it simply, organic wines are made from grapes that are organically grown, without the use of artificial fertilisers or synthetic pesticides, including fungicides, herbicides, soil fumigants, growth regulators or hormones. In general, organic methods focus on prevention rather than cure and embrace quality, purity, safety and health. By working with nature, organic methods safeguard the long-term well-being of the environment and the consumer.

In order to achieve this, the main emphasis is on building a living soil and environment that encourages beneficial organisms at all levels. When the land is enriched, the soil creates strong, healthy crops. These plants have more inherent resistance to diseases and pests and therefore need less intervention – chemical or otherwise. This focus in the vineyard on a balanced and healthy plant distils down into top-quality fruit. Organic grapes should provide the best raw material to quality wine – and with excellent raw material to work with, the need for chemical manipulation in the winery decreases. The ultimate goal is therefore to produce a quality wine that is grown and produced in as natural a way as possible, without causing damage to the surrounding environment.

11.3.1 *Soil*

Organic viticulture begins with the minutiae of the soil, the tiny microbes and bacteria that work in living soil to release nutrients on which plants feed. There are as many as five billion to a teaspoon, working away to create living soil that feeds the plants that feed the world. Artificial fertilisers and synthetic pesticides, which are poisons, act on beneficial and harmful micro-organisms alike, rendering much conventional land 'dead'. Organic methods are in direct contrast to this, encouraging equilibrium with the land.

In the vineyard, nitrogen must be fixed into the soil naturally. This is done by planting leguminous crops such as clovers or beans between the vines. These plants have bacteria living on their roots that fix nitrogen from the air. When the plants are dead, they are ploughed in, decay and release nitrogen into the soil. At some wineries, commercial crops such as strawberries are rotated in, which add to the profitability of the vineyard.

The soil is maintained and improved with organic compost and fertilisers which can be made from a range of materials including animal manures and waste products from the fermentation, which are usually composted and spread on the ground under the vines. Typical composting material includes yeast deposits, sediment, marc, vine leaves, prunings and straw.

Earthworms and other such creatures break this matter down in the first stages. Then billions of micro-organisms break that matter down further into basic nutrients that support and feed the vines. The roots of the vines search deep into the ground to draw the natural fertilisers up, creating a strong plant. Natural minerals are allowed to be added to the soil, according to the regulations, to prevent nutrient shortages or imbalances. Soil analyses provide important information about soil acidity/alkalinity (pH) and corrective measures that may need to be taken.

Traditionally, organic winemakers have chosen grapes to suit their soil types and have chosen varieties for their character and their resistance to disease, rather than for their ability to produce high yields. This is true in other organic agriculture where farmers plant arable crops that produce lower yields but are able to withstand pests, diseases and falling over.

11.3.2 *Pests and diseases*

Organic methods differ radically from conventional when it comes to pests and diseases. As well as the focus on a stronger and healthier plant that can withstand these problems, the organic grower can use other techniques in and around the vineyard. In some areas of France, organic farmers find little problem with pests, which is comparable to organic arable farmers who learn not to panic at an attack of blackfly. Although pests do afflict the organic vineyard, and mildew has no cure, most organic wine producers are not affected by either to the point where they cannot run a commercially successful operation.

Interplanting (Fig. 11.1) is a technique which combats the pest problem in a few ways. Firstly, it brings the pests' natural enemies into the vineyard. This can be done by planting flowers, herbs and crops in between the rows that attract mammals, birds and insects. Predators such as ladybird, hoverflies and parasitic wasps remain in the crops between the vines until a pest starts to multiply. They then emerge and attack the pests, breeding and multiplying at a pace that will stop the pest. Another method is to interplant decoy crops which vine pests prefer. The broader picture of interplanting is that it reduces the impact of perennial crops by creating a type of rotation or biodiversity, resulting in a more balanced environment against pests and diseases.

Organic growers also use biological control. They are allowed to import natural enemies of pests. This works either through predation or parasitism. This has proven effective for the control of red spider mite and specific caterpillars, and does not affect other insects. And a third technique is to use insect traps ranging from sticky yellow strips to pheromone or sexual lures. The pheromone traps use synthetic extracts of the chemical scents that

Fig. 11.1 Interplanting to attract beneficial insects.

many insects emit when attracting mates. They are species specific and can be hung throughout a vineyard.

As a last resort, if pests get of control, organic growers are permitted to use plant, herbal or mineral-based sprays although, in general, spraying is discouraged. There is an insecticide made from chrysanthemum that is frequently permitted in organic wine growing and other agricultural systems. Rotenone, from the tropical plant derris, is often permitted, as well. These two approved insecticides break down in the environment quite quickly and are used against aphids and caterpillars. Also, the organic movement is dynamic rather than static and new research is continually being carried out to find new ways of handling pests without damaging our beneficial insects and pollinators.

There is no cure for rot and mould, but instead of relying solely on sprays, organic farmers practice canopy management, training the vines in a more open style to maximise leaf exposure and the circulation of air. This will reduce the risk of rot in combination with minimal spraying – and stresses the vine, encouraging high-quality grapes. The traditional Bordeaux mixture (lime and copper sulphate), which has been used for centuries as a fungicide, is allowed in organic vine growing although the frequency of spraying is greatly reduced for an organic grower. Conventional vineyards can be sprayed around 15 times a year, compared to an organic vineyard where spraying is severely limited. Bordeaux mixture and sulphur are also used in preference over systemic fungicides that are taken up into every part of the plant and remain in the plant. The Bordeaux spray is topical – and has restricted use near the time of harvest, to reduce the risk of chemical contamination in the finished product.

11.3.3 *Weeds*

There are no organic weedkillers, so the organic grower must adopt an entirely different approach. And it is important to note that it is not the intention to completely wipe out all weeds in the vineyard. Weeds are used to make compost and can therefore be beneficial, but they must be prevented from competing with the vines for water and nutrients in the soil. The three techniques used are weeding by hand, weeding by machine or covering weeds with mulch to smother them. Organic vineyards and farms often exhibit creative use of machinery – either modern equipment adapted for new uses or old equipment brought back into use to control weeds as they did 40 years ago.

11.3.4 *In the winery*

Turning grapes into organic wine follows the same principles as conventional wine. The grapes are picked, pressed, fermented, filtered and bottled. As any winemaker knows, however, this does not begin to describe the complex affair of making a good or even a great wine. Today wine is the result of a highly technological processing operation, which for the conventional winemaker includes a wide array of different additives. Here again, the organic producer departs from his or her conventional counterpart because the extent of manipulation of the raw material and the use of additives is regulated and controlled. The organic regulations on additives are essentially practical, aiming to reduce inputs without crippling the producer. Reducing the chemicals allowed in the winemaking process produces an individual and distinctive wine. The wine is more genuine because it has not been tampered with to the same degree as conventional wine.

Because of reduced spraying, organic grapes coming into the winery still retain their natural yeasts. The organic winemaker normally ferments the wine using these wild yeasts, following the most natural process and ensuring a wine with individuality.

Although maximum amounts vary from country to country and region to region as well as from different types of wine, the goal is to reduce sulphur levels to create a healthier product that is still stable. Sulphur dioxide is allowed in very restricted amounts, the Directive 87/822/EEC setting maximum levels as shown in Table 11.1.

The total sulphur is the amount added initially and the free is the amount that remains in the wine. Without a doubt, the lower the sulphur dioxide, the better it is for an individual's health. Most organic winemakers use very little sulphur and some use none at all. In the Southern Rhône, Domaine Gramenon bottles some of its wines with no sulphur. The Frey vineyard in the Redwood Valley of California does not use sulphur in its winemaking either, which has not shortened the life of its wines so far. Frey's 1988 Organic Syrah was described in *Wine* magazine as 'still a very lively wine with full rich flavours of plums, liquorice and leather with a fading tannic back-up' (Alloway 1999). One of the owners feels that cleaner winemaking and the use of stainless steel has negated the need for sulphur. Not everyone will agree, but is interesting to see stable, successful wines ageing and emerging that are made without it.

According to *E for Additives*, sulphur dioxide (E220) occurs naturally but is produced chemically by the combustion of sulphur or gypsum:

'One of the oldest food additives known to man, sulphur dioxide was employed by the Romans, Ancient Greeks and Egyptians as a preservative for wine. Today it is the most reactive food additive in use and one of the most versatile…

'When sulphur dioxide dissolves, the disulphide chemical bonds which result destroy the vitamin B_1 or thiamine in foods by breaking up the protein molecules. Sulphurous acid, produced when sulphur dioxide is dissolved, may cause gastric irritation. Healthy people have no problem metabolising sulphur dioxide: the kidneys and liver both produce enzymes which oxidise sulphites, but those with impaired kidney and liver may need to avoid sulphites. Foods containing sulphites may precipitate an asthmatic attack in asthma sufferers, who are very sensitive to the irritant effects of sulphur dioxide gas which may be liberated from the foods containing it and inhaled as the food is swallowed. It is one of the additives which the Hyperactive Children's Support Group recommends is eliminated from the diets of the children it represents.'

Table 11.1 Maximum permitted levels of sulphur dioxide (total and free) in organic wine (source: Directive 87/822/EEC).

Wine	SO_2 total (mg/l)	SO_2 free (mg/l)
Red	90	25
White	100	30
Rosé	100	30
Cider	100	30
Sparkling	100	10
Dessert	250	70

Once the fermentation process is completed, the wine is usually clarified by fining, filtering or centrifuging. The most common fining agents in organic production are egg whites and bentonite or kieselguhr clays. Several certifying bodies also permit casein, isinglass and gelatine, but these are used less often. Filtering, which removes unwanted particles, also strips the wine of some of the enzymes that give wine flavour and character. Because of this, some of the best organic winemakers neither fine nor filter, but leave the wine in the cellar and allow the action of time to settle the particles and produce a clear wine. This is not always financially viable for winemakers, and clearly white wines that are to be sold the year after harvesting will need to be filtered.

One of the most salient points about organic winemaking is the parallel between them and the makers of fine wines in general. Both groups follow the guidelines below:

- Produce wines with a minimum of added permitted chemicals
- Use lower levels of sulphur dioxide
- Know that the best way to improve quality is to decrease yields
- Emphasise quality over quantity
- Strive for individuality rather than uniformity
- Combine the best traditional methods with the benefits of modern technology.

They also share the knowledge that high-quality grapes combined with high-quality winemaking produces high-quality wine. Great grapes can be turned into poor-quality wine, but poor-quality grapes will never be turned into great wine. To this end, it tends to be the cheaper end of the wine market that suffers from the use of too much sulphur, where there can be poor quality control and less than excellent hygiene standards. Many of the fine winemakers of the world do not use artificial fertilisers because, by keeping yields down, quality improves. They follow organic principles to create excellent grapes and wines but may not even realise it and are not certified organic.

11.4 Organic Grapes into Wine Alliance (OGWA)

A look at a specific certifying body, Organic Grapes into Wine Alliance of California, USA, will show the sorts of rules and guidelines that are laid down for the organic winemaker. The OGWA was established in 1989 to support the production of wines made from organically grown grapes, and has established the production standards shown in Table 11.2.

11.5 Fetzer spotlight

These regulations set out what a winemaker can do, cannot do and should do in the vineyard and in the winery. The obvious gap in the regulations of all certifying standards for someone wishing to convert is *how* to achieve what is being specified. Organic winemakers to date have largely been pioneers, finding their way by trial and error. We now take a look at a major winery and its path to organic winemaking, to understand the methods used in a profitable, successful winery and give potential converters confidence and a clear route forward.

Table 11.2 Production standards of Organic Grapes into Wine Alliance of California.

Section	Standard
Section I: Grape Origin	*Recommended*: third party certified organically grown wine grapes. *Tolerated*: wine grapes grown according to California organic standards, but not third party certified. We suggest that each OGWA member winery be responsible for verification of the cultural practices of their wine grape sources. *Prohibited*: wine grapes grown with synthetic herbicides, pesticides, fungicides or fertilisers
Section II: Harvest Crushing	*Recommended*: select only superior fruit and avoid rot or mildew. Pick into shallow boxes to avoid breaking grape berries. Transport to winery as soon as possible in gondolas or boxes. Separating press fractions and using only highest quality juice or wine. Judicious cleaning of gondolas and equipment between uses and intensive cleaning nightly during harvest season. *Tolerated*: mechanical harvesting equipment. Enamel-lined gondolas (food grade enamel kept in good condition)
Section III: Yeasts	*Recommended*: active yeast cultures and yeasts present in the must. Organic yeast nutrients when nutrients must be used. *Tolerated*: yeast additions. *Prohibited*: inorganic yeast nutrients
Section IV: Sulphur Treatments	SO_2 levels in California organic wines shall be no more than 100 parts per million total and 30 parts per million free SO_2 at the time the wine is released. It is always recommended that winemakers strive to use as little SO_2 as possible. *Accepted*: solutions of greater than 5% SO_2 up to saturation, prepared on premises by bubbling gas through water. Accepted times of SO_2 application: (1) when cleaning barrels; (2) at bottling; (3) upon completion of fermentation (It is tolerated, but not advised, to apply SO_2 during crushing or pressing, or during ageing on the lees). *Tolerated*: sulphur wicks on cellulose supports in barrels. Use of SO_2 gas in maintaining empty cooperage. *Prohibited*: asbestos wicks. Potassium metabisulfite
Section V: Stabilising Agents	*Recommended*: none are recommended. *Tolerated*: citric, tartaric, malic, ascorbic, fumaric acids from non-synthetic sources according to BATF standards. Low temperatures for tartrate stabilisation (cold stabilisation). Flash pasteurisation with technical justification. *Prohibited*: potassium ferrocyanide, synthetic citric acid, metatartaric acid, sorbic acid and sorbates
Section VI: Clarification/Fining	*Recommended*: natural settling and racking. *Tolerated Clarifying Materials*: fish based fining agents. Non-hydrolysed bone gelatine. Bentonite. Kaolin. Pure casein, guaranteed free of residue. Diatomaceous earth. Fresh egg whites. *Tolerated Clarifying Processes*: cellulose plate filters. Centrifugation. Sterile filtration using membrane filters. Cross-flow filtration with FDA approved materials. *Prohibited*: hydrolysed gelatine. Asbestos filters
Section VII: Colouring/De-colouring	*Recommended*: no colouring or de-colouring agents are recommended. *Tolerated*: natural active carbon. *Prohibited*: all colouring agents and carbon black from incomplete combustion of combustible fossils

Table 11.2   (*Continued.*)

Section	Standard
Section VIII: Volatile Acidity	*Recommended limits*: current BATF standards *Prohibited*: above BATF limits
Section IX: Acidification/ De-acidification	*Recommended*: an earlier grape harvest, higher acid fruit or blending higher acid wine for acidification. Malolactic fermentation for de-acidification *Tolerated*: tartaric, malic, citric and tumaric acids (from natural sources if suppliers exist) according to BATF regulations. Calcium carbonate (max. 75 p.p.m.) and cream of tartar from natural sources, if they exist
Section X: Storage Vessels	*Recommended*: wood barrels and tanks (kept full). All containers be kept as full as possible to minimise contamination. Inert gas should be used to fill any space not occupied by wine. Stainless steel tanks and containers if cleaned in accordance with California State law *Tolerated*: certain plastic materials if they meet State standards for potable water containers. Plastic lining in grape bins. Food-grade silicon bungs in wood barrels
Section XI: Transportation of Bulk Juice and Wine	*Tolerated*: polyethylene containers if they meet potable water standards. Stainless steel tanker trucks if cleaned according to DHS standards. Wooden cooperage
Section XII: Bottling/Packaging:	Glass bottles are mandatory *Recommended*: sparging bottles with inert gas before filling. Recycled glass *Prohibited*: plastic bottles or containers and cans. After 1/1/92, all lead, chromium, mercury and cadmium will be eliminated from packaging materials
Section XIII: Corking	*Recommended*: high quality, natural cork *Tolerated*: particle cork glued only with high purity elastomere resins excluding all solvents, plastisizing agents and formol. Plastic lined crown caps during sparkling wine processing. Natural cork treated with chlorine and SO_2 *Prohibited*: composite cork, polyurethane, solvents and plastisizing agents. Plastic corks. Corks treated with fungicide and pesticide
Section XIV: Cleaning Agents	*Recommended*: the use of cleaning agents as permitted by California State Health & Safety Code, followed by adequate water rinsing before vessels come in contact with wine

Organic wine cannot be called a 'hippy' product. Firstly, hippies do not necessarily drink a lot of wine, and secondly, mature companies like Fetzer Vineyards in California produce excellent quality organic wines that are taking awards alongside some of the world's finest. Today, with 700 acres, Fetzer Vineyards is one of the largest certified organic vineyard holdings in the world. The climate in the region is especially suited to organic viticulture: cold winters and dry summers mean that downy mildew is unknown and grey rot is rare.

Fetzer Vineyards has it roots back in America's history some 40 years ago when lumber merchant Barney Fetzer bought a ramshackle property, Home Ranch, in Redwood Valley, Mendocino County, California, 105 miles north of San Francisco. Although the house's main function was home to his wife, Kathleen, and their 11 children, it came with 720 acres of land for sheep, apples and pears, hunting and fishing. More interestingly, a 100-year-old vineyard was included.

Barney and his sons started uprooting the old vines and replanting with Cabernet Sauvignon and the native American Zinfandel. They harvested the grapes and sold them on to small home winemakers throughout the country. At some point along the way, they decided to make their own wine, and in 1968, two of the sons, Jim and John, converted an old barn into a winery. That year the Fetzer family produced 2500 cases of their first batch of wine. Barney's motto was to 'make wines people can drink every night'. Their hefty reds soon earned them a reputation for quality wines at reasonable prices.

Although the family's attempts at white wine in the form of Chardonnay proved unsuccessful, they carried on making wine through to 1976, when, with sales of red wine booming, Barney retired from the lumber business and dedicated all his time to making wine. The year after, Fetzer hired Paul Dolan as head winemaker, a recent graduate from the University of California, Fresno, with a master's degree in Oenology, and gave him the task of establishing Fetzer's first white wines.

Over the next several years, Fetzer earned a reputation with its whites, primarily Chardonnay and Gewürztraminer, purchasing further acreage and surviving the death of Barney in 1981 after which ten of his children took over the management of the company. By the mid-1980s, the Fetzers, like many premium California wineries, were trying to link their wines to foods. To this end, Fetzer established the Bonterra garden, a 5-acre organic garden at Valley Oaks, under the guidance of a master gardener with an extensive knowledge of organic growing methods (Fig. 11.2). This garden is a work of art, alive with colour, insects and a huge range of fantastically flavoured fruits, vegetables, edible flowers and herbs. It was the quality and taste of the produce from this garden, combined with a

Fig. 11.2 Organic gardening at the Bonterra Garden where organics began.

concern for conservation of land and water and the safety of their vineyard workers, that encouraged Fetzer to experiment with organic viticulture.

In 1988 they converted the 131-acre Home Ranch vineyard to organic vine growing. They were so impressed with the results that in 1990 Fetzer made the enormous commitment to grow all of its grapes organically and launched into a 5-year conversion period. Fetzer Vineyards became the first winery in California to make such a commitment to organic viticulture and today all of its vineyards are certified organic by the California Certified Organic Farmers (CCOF). They claim that ultimately, organic farming is less expensive and that the initial investment in additional equipment such as compost spreaders, weeding machines and cover-crop seeders is recouped within two years.

The general consensus in the industry is that the majority of chemical agricultural aids were developed after World War II. Fetzers confirm that before 1945, farmers had only non-synthetic means of growing crops and that pest and disease management was part of a larger strategy, not a quick fix from chemical compounds. The company has now gone back to traditional methods that include planting cover crops between the rows of vines. This creates a dense carpet that attracts beneficial insects, prevents erosion and provides green manures when they are ploughed back into the soil to release nitrogen. It firmly believes that farming organically is about preventing problems, not treating them after they have arisen. As part of its research and pioneering in organic viticulture, the Fetzers maintain a 10-acre experimental vineyard where they trial vine spacing, trellising, pruning, rootstocks, varietals and clones.

Other environmental strategies include eliminating lead capsules by replacing them with a beeswax stopper. Fetzer's organic range of wines called Bonterra is packaged in recycled glass bottles, with labels made out of kenaf (paperless hemp), printed with soy-based ink.

Pests

Pests are controlled by actively encouraging, releasing and monitoring beneficial insects and organisms. The Fetzers plant plum trees all around the vineyards to attract the tiny parasitic wasp *Anagrus* that feeds on leafhoppers. Companion plants such as mustard, crimson clover and sunflowers are planted in rows alternating with permanent covers in the other rows across the entire vineyard to attract other beneficial insects. And ladybirds are introduced to feed on aphids.

Soil

For the soil, the Fetzers plant legumes like bell beans and crimson clovers to build nitrogen. To add humus to the soil, they also compost the grapes after harvesting to spread throughout the vineyards. Over a 2-year period, the Fetzers converted all of the stems and seeds from 42 000 tons of crushed grapes into compost. The Fetzers also plough in cover crops such as rye, vetch and oats for humus. Other organic wineries use fodder radish for leaf humus and manure. Winter rye not only adds humus, but also improves the soil structure with its fibrous roots, and vetch is excellent for providing nitrogen. In Australia, organic vineyards plant lucerne between the rows because its enormous root system aids in the formation of good soil structure. To help aerate the soil, the Fetzers grow dycon radishes that have a long tap root and penetrate deep into the ground.

Weeds

For handling weeds, the Fetzers have developed specific machines that go in and out of the vines, again the focus being on controlling weeds, not eradicating them. Other wineries work the ground manually to keep weeds under control.

Diseases

Moisture trapped in a dense canopy will create an ideal environment for mildew and rot. The trellis system used by Fetzer aims to keep the fruit and canopy away from each other. They use a vertical trellis, bilateral cordon trained, with spur pruning, the canes being grown vertically (Fig. 11.3). They then have what they call a 'fruit zone' with a canopy of leaves overhead. This canopy is thinned by hand at various times during the growing season to provide aeration and allow sunlight onto the grapes.

Phylloxera has recently been destroying vines across California, but has not troubled the Fetzer vines to date. Although phylloxera is in neighbouring vineyards, the Fetzers think that their vines have not been affected because of the inherent strength and health of the vines as a result of organic practices. There is also concern that Pierce's disease, transmitted by a small insect called a sharpshooter that carries the bacterial disease in its mouth, could become as big as problem as phylloxera. To this end, Fetzer have left the bug's natural habitat of trees and hedges surrounding the vineyard and hope this will negate the need for them to venture into the vineyard. If phylloxera and/or Pierce's disease become a problem, however, they will have to replant with resistant rootstock.

Fig. 11.3 Vertical trellis training to avoid mildew and rot.

Awards

All of these techniques have combined to create top-quality grapes. The organic Bonterra Vineyards range scooped four silver awards in *WINE* Magazine's International *WINE Challenge* 1999 with their 1996 Chardonnay, 1997 Cabernet Sauvignon, 1998 Viognier and 1997 Zinfandel and a bronze with the 1997 Syrah (Fig. 11.4). This shows that organic winemaking is viable and that the wines can take their place among some of the world's finest.

11.6 Biodynamic

According to a 1998 issue of WINE magazine, Claude Bourguignon of the *Laboratoire d'Analyse Microbiologique des Sols* believes that modern vineyard techniques produced soils in parts of the Côte d'Or which have 'less microbial life than the Sahara'. While winemakers such as Fetzer have turned to organics to combat the ravages of large-scale chemicals and mechanised harvesting that have killed and compacted the earth, other winemakers have skipped straight to biodynamic winegrowing – unusual because it is the extreme of organic which combines science with spiritual practice. In winemaking, biodynamics follows some organic principles in terms of enriching the soil, but has an added dimension, sometimes described as mixing in elements of astrology and homeopathy. It believes in the interrelationships of all kingdoms – mineral, plant, animal and human – and their intricate correspondence to the larger cosmos.

Fig. 11.4　The award-winning range of Bonterra organic wines.

In the standards, biodynamics forbids the use of chemicals and fertilisers except for Bordeaux mixture. Growers rely on herb-based compost and field sprays, including extracts of camomile, nettle, oak and valerian. Composting is a key activity, but it is not just about returning nutrients to the soil as in organics. Biodynamic composting regards vegetable waste, manure, leaves and food scraps as all containing precious vitality as well, which if handled properly, can be transferred into the soil. Cows' horns filled with dung are buried in the soil in the belief that they will maintain the harmony and balance of the soil and plants. And there are biodynamic preparations used directly in the field – one on the soil before planting and one on the leaves of growing plants – the effects of which have been verified scientifically. The vineyard and winery operations are governed by the position of the planets and the phases of the moon, the *Sowing Calendar* guiding growers towards the time to plant or prune vines or when to harvest the grapes.

These practices are not all as odd as they may sound at first. Long before Steiner developed his anthroposophy, peasant farmers in Europe consulted the moon. Everyone agrees that the phases of the moon affect the times of racking and bottling. Wines open and close in the bottle and become more or less expressive when nosed also according to the phases of the moon – because of the corresponding changes in atmospheric pressure that affects the placidity of the wine. According to *The Wild Bunch*, Telmo Rodriguez of the Remelluri Estate in Rioja, Spain, prune their young vines during a waxing moon and their old vines during a waning moon according to tradition, not to biodynamics. In France, Patrick Doche, the owner of Chateau Cayla in the Entre Deux Mers says, 'the moon has such a big influence on wine. It's absolutely correct that you shouldn't rack either on a full moon or a new moon. If you do you destroy the wine.' The forces of the moon are widely studied and understood in France, particularly in Burgundy.

Well-known, famous and world-class wineries grow their vines biodynamically, sometimes following the Steiner guidelines to the letter, sometimes following certain practices and avoiding the more extreme mystic aspects. Anne-Claude Leflaive, manager and part owner of Domaine Leflaive in Puligny-Montrachet, has no doubts that biodynamic methods have improved the quality of her vines and wines. The domaine's reputation slid in the late 1980s, which Anne-Claude thinks was attributable to the increasingly inert vineyards. In 1990 she converted one hectare to biodynamic agriculture, extending it through the decade, until today the entire 22-ha domaine is managed to these standards. Anne-Claude believes that biodynamic techniques have increased microbiological life in the soil and lowered the incidence of rot.

André Ostertag in Alsace thinks that biodynamic winegrowing has brought about astonishing changes in his vines. After two years of biodynamic growing and painstakingly thinning the leaves by hand on his squat and bushy vines, they have recently started growing in a different style, taller and more open so that less pruning is needed. Two notable names from the Loire – Noel Pinguet of Domaine Huet and Nicolas Joly of Clos de la Coulée de Serrant – are also Steiner followers, as well as Robert Eden in Languedoc-Roussillon with his stunning biodynamic Comte Cathare range of wines. Other advocates include followers in Bordeaux, Champagne, California and New Zealand.

With its celestial influences and herbal infusions, biodynamic may not be for everyone and not every follower wants to shout it from the label. There is resistance to the methods, probably because they are difficult to understand, impossible to prove and rely to a degree on faith. Nicolas Thienpont bought Chateau Pavie-Macquin in St Emilion in 1994, leaving a quarter of the land under biodynamic practices and resuming the use of synthetic chemicals on the rest. He has found, however, that the 10 biodynamic acres use only one-third of the quantities of copper and sulphur treatments usually prescribed – and that they have flourished. He is returning the remainder of the vineyards to biodynamic ways … on the quiet.

Is biodynamics worth the extra effort, unquestioning faith and possible ridicule? According to some, the answer is a resounding yes. In *The Wild Bunch*, Robert Eden says that 'unlike organic culture, the culture of biodynamics is very disciplined – it has its rules and regulations and if you keep to them you really feel the difference'. The brothers at the House of Chapoutier in the Rhône Valley insist that biodynamics has transformed their vineyard. Floods in 1993 devastated much of the year's vintage, including up to 90% of the Hermitage grapes but only 10% of the Chapoutier grapes were lost. The brothers attribute this to the biodynamic methods that cause the roots to grow much deeper and avoid damage. The same could, of course, be said of organic, but it is in the soil and the tasting that the difference really shows.

Soil microbiologist Claude Bourguignon did comparative analyses of organic and biodynamic vineyards and found that for one class of bacterial micro-organism, there were 400 per gram in the organic soil, but 100 000 per gram in the biodynamic soil. Anne-Claude at Domaine Leflaive has been comparing organic and biodynamic wines from Clavoillon since 1990 when the first acre was converted. The grapes are picked at the same time and vinified in the same way. She says that, until 1995, the differences between the two were not always apparent and that the organic wines sometimes came out on top. But since 1995, when the effects of biodynamic methods had become established in the vines, the biodynamic wines win at blind tastings every time.

11.7 EU labelling regulation

Wander down any supermarket aisle and you will see the occasional person looking at a label. Wander through an organic shop and you will see several people staring at labels to determine what is in the food they are buying. Artificial additives and preservatives, refined sugar, hydrogenated fat and gluten are some of the ingredients that the educated consumer is looking out for and trying to avoid. Pick a bottle of wine off of the shelf and amazingly, there are no ingredients – not grapes or sulphur dioxide or yeast or a whole host of additives allowed in the production process. If wine labels did display the ingredients used in their production, organic wines would have an edge over conventional wines.

The Food Labelling Regulations of 1984 introduced the E-code, which made it easier to identify some of the additives in our food, but any drink with an alcoholic strength by volume of more than 1.2% does not have to list the ingredients. This, by implication, means that it is actually against the law to print ingredients on a bottle of wine. All other food and drink has a list of ingredients so that consumers – those who are interested and those who have allergies – can see what is in the food they are buying. Why alcoholic beverages are excluded from this Regulation is unclear. Some speculate that it is the result of powerful beer and wine lobbies in Brussels. One can only wonder.

In America and Australia, wines have to declare sulphur dioxide on the bottle, but still do not give full disclosure of other ingredients. In the US this reads as 'contains sulphites' and in Australia 'contains Preservative E220', but this regulation only applies to wines consumed in their own countries. Australian wines also list antioxidants such as E300 (ascorbic acid or vitamin C). And US wines must carry a health warning that reads as follows: 'Government Warning: According to the Surgeon General, women should not drink alcoholic beverages during pregnancy because of the risk of birth defects. Consumption of alcoholic beverages impairs your ability to drive a car or operate machinery, and may cause health problems.' These wines must be relabelled when they come into the EU.

Ideally, all wines should list any additives used during the winemaking process. But in the *Official Journal of the European Communities*, 'Council Regulation (EEC) No. 2392/89 of 24 July 1989 laying down general rules for the description and presentation of wines and grape musts' 27 pages of rules are laid down concerning what is allowed on wine labels. In the 'description of quality produced in specified regions', Article 11 sets out that the description should include the following information:

- The name of the specified region of origin
- The nominal volume of the wine
- The name of the bottler and local area and member state
- The alcoholic strength by volume.

The description on the label *may* be supplemented by the following information:

- A statement as to whether the wine is red, white or rosé
- The vintage year
- A brand name
- The name or business names of the persons involved in the distribution
- A distinction which is likely to enhance the reputation of the wine
- Certain analytical data other than the alcoholic strength by volume
- A recommendation to the consumer as to the use of the wine
- Additional details of a traditional kind
- The Community expression 'quality wine produced in a specified region' or 'quality wine psr' or a traditional expression
- Details as to the method of production, the type of product, the particular colour of the quality wine
- The name of a geographical unit that is smaller than the specified region
- The name of the vineyard or group of vineyards where the quality wine was made
- The name of one or two vine varieties
- A quality control number allotted by an official body
- An award granted by an official body
- A statement that the wines were bottled either at the vineyard where the grapes used were harvested and made into wine, or by a group of vineyards, or in an undertaking situated in the specified region indicated or the immediate vicinity of that region
- Information in respect of bottling in a specified region
- The number of the container or the number of the lot
- Information concerning the history of the wine in question, the natural or technical conditions governing the production or the ageing of the wine
- The lower-case letter 'e', indicating that the pre-packages satisfy the condition laid down in Direct 75/106/EEC as regard filling.

The question is: what information are today's consumers interested in? With the increase of allergies and asthma, and increased awareness of additives in foods, one would expect that people would be more interested in additives than awards. Article 12, however, states that 'only the information specified in Article 11 shall be allowed for the description on the label of a quality wine', so the consumer is kept in the dark.

11.8   Vegetarian and vegan

Certain people may object to the finings used in wines on moral grounds, but the lack of relevant additives listed on the label make it impossible for the consumer to be able to choose. Permitted finings include isinglass (from the swim bladder of certain tropical fish, especially the Chinese sturgeon), egg albumen, gelatine (from animal bones), modified casein (from milk), tannin (from wood), chitin (from the shells of crabs or lobsters) or ox blood (rarely used today). Non-animal alternatives include bentonite, kieselguhr and kaolin clays, and silica gel or solution.

The use of animal derived products in the production of alcoholic beverages in general is fairly widespread and, although fining agents are removed at the end of the winemaking process with the possible exception of minute quantities, this is not the case in cask ales. Isinglass is usually used to fine cask-conditioned ales to clear the material, especially the yeast, held in suspension in the liquid. This then sits in the bottom of the cask, and if disturbed, can be seen in a pint of real ale held up to the light as cloudy lumps swirling around. These finings have always been used and some say there is little demand from the consumer for an alternative, probably because the consumer is largely ignorant of these processes (which are not mentioned on labelling).

With the growth of the organic industry, consumers are becoming more educated and more interested generally about food and the chemicals used in its growth and production. Retail specialists of organic wines have more and more consumers asking for vegetarian and vegan wines and the supermarkets are being questioned as well. There is a misconception among some consumers that organic wine equates to vegetarian wine and this is not always the case. Although the majority of organic wines are vegetarian or vegan, it is not a requirement.

Because most vegetarians consume dairy products and eggs, casein and egg albumen are acceptable to them as finings, providing the eggs are free-range. Vegan wine would need to be fined with clays, or using new methods such as centrifuging or filtering. Some organic wine producers rely on time in a cold cellar to allow the wine to settle and therefore use no finings, however this is only practical for small-scale production and would not suit huge companies producing large quantities of wine for cheap markets.

11.9   Beer and the rest

Organic beers, including wheat and hemp beer, are produced on the Continent and in the UK. In England's market of declining beer sales and closing breweries, there is the organic Golden Promise from the Edinburgh based Caledonian Brewery, and Organic Best Ale from Samuel Smith of Tadcaster in Yorkshire. Between these makers, they buy up all the organic English malt and find it difficult to secure enough organic hops, the former buying the entire crop of commercially grown British hops and the latter importing organic hops from New Zealand.

Peter Hall, in Kent, seems to be the only hop grower in the UK with an interest in organics. He plants and ploughs in legumes for nitrogen and to conserve nutrients in the soil; he sows white mustard seed between the trellises to create a habitat for lacewing and ladybird to attack the damson aphid; and he sprays the tops of the bines with a soapy water solution to kill aphids. Mildew, which can blacken hops, is handled with a combination of spraying with copper oxychloride (the only Soil Association-approved fungicide) and digging out infected bines.

In Germany, two large organic certification bodies, Naturland and Bioland, encourage the growth of organic barleys and hops and an area of the Hallertau, the world's largest hop-growing region, will soon be planted with organic hops – a great sign for the future of organic beer. The organic Pinkus ales, wheat beers and lager are brewed by one of Germany's best-known brewers. In France, one of the top brewers in Castelain is now brewing Jade organically, as is Dupont in Belgium, with its two organic Saison ales.

Organic alcoholic drinks apart from wine and beer make up the minority. Beyond the fair selection of beer, there are smatterings of other drinks that tend to be found in specialist organic retail outlets. These include organic perry, mead, cognac, cider, sherry, port, sake, whisky, crèmes and vodka. But as more alcoholic drinks come onto the market, the specialist retailer is able to offer a very comprehensive range to the discerning consumer.

11.10 Sellers

The market for organic alcoholic drinks is growing, marked in the UK by recent openings of organic restaurants, pubs and bistros as well as established restaurants and hotels turning to organics for quality. The big supermarkets are also selling organic drinks, and some have a commitment to increase their ranges. The three largest supermarkets in the UK sell organic beverages as follows (at the time of writing):

- Sainsburys: one white wine and two red wines with the intention to expand the range
- Tesco: one wine and one beer with the possibility of extending the range
- Waitrose: three to four white wines, eight red wines plus beer and cider and always keeping an eye on the organic alcoholic beverages available.

Then there are specialist organic supermarkets like Planet Organic, with over 200 organic alcoholic beverages and another specialist organic retailer, Bumblebee, with the same. Although Bumblebee is a single shop operation, Planet Organic plan nationwide expansion with 35–40 stores across the country over the next 10 years. Other organic retailers, which offer a much smaller range of wines, are also planning on expanding, illustrating the potential growth for organic wine sales in the UK.

11.11 Associations

The first step for producers interested in making organic wine is to join an independent organic association in their area. This group will support the winemaker through the process of conversion and set controls that define everything from how the grapes are grown to how the wine is made and what substances are allowed. Once the regulations are followed, the winemaker can display the symbol of the association on the label. As the organic wine movement grows, a clearly defined common standard will, it is hoped, evolve, accepted by the various certification bodies across the world.

In each wine producing country, there are several organic organisations. Table 11.3 lists the largest in their respective countries, which will hold details of other regional bodies.

References and bibliography

Alloway, K. (Mar 1999) Organic Chemistry. *WINE.*
Brook, S. (Sep 1998) Biodynamic Woman. *Decanter.*
Brown, L. (1998) *The Shopper's Guide to Organic Food.* Fourth Estate Ltd, London.
Carson, R. (1962) *Silent Spring.* Penguin Books, London.
Fetzer Vineyard (1998) *Fetzer Vineyard History.*
George, R. (Mar 1993) Red, White & Green. *Country Living.*
Hanssen, M. with Marsden, J. (1987) *E for Additives.* Thorsons, London.
Joseph, R. (2 Mar 1997) Pure pressure. *Sunday Telegraph Magazine.*
Lockspeiser, J. & Gear, J. (1991) *Thorsons Organic Wines Guide.* Thorsons, London.
Matthews, P. (1997) *The Wild Bunch.* Faber & Faber Ltd, London.
Matthews, P. (Nov 1997) Star Treatment. *Food Illustrated.*
Official Journal of the European Communities (August 1989) Council Regulation (EEC) No 2392/89 of 24 July 1989.
Orr, C. (Jun 1999) The Garden of Eden. *WINE.*
Protz, R. (10 Oct 1998) Hops Away. *Observer Magazine.*

Table 11.3   Organic wine organisations worldwide.

Country	Organisation
Australia	National Association for Sustainable Agriculture
France	Nature et Progrès
	Union Nationale Interprofessionelle de l'Agrobiologie
	Fédération Européenne des Syndicats d'Agrobiologistes
Germany	Bundesverband Okologischer Weinbau
Holland	EKO – Stichting Ekomerk Control
Italy	Associazione Italiana per l'Agricultura Biologica
New Zealand	New Zealand Biological Producers' Council
Portugal	Associacao Portuguesa de Agricultura Biológica
Spain	Vida Sana
	Umbella
UK	UKROFS
	Soil Association
USA	California Certified Organic Growers

Rand, M. (Sept 1998) Stars in Their Eyes. *WINE*.

The Soil Association Organic Marketing Co Ltd (Mar 1999) *Standards for Organic Food and Farming*

Hall, E. (Jan 1992) So green, so what? *Which? Wine Monthly*

www.isgnet.com (27 Jul 1999) *Standards for Wines Produced from Organically Grown Grapes*

www.demeter-usa.org (27 Jun 1999) *What is Biodynamic Agriculture?*

www.vegsoc.org (7 Mar 1999*) The Grape Divide*

www.vegsoc.org   (7 Mar 1999) *Vegetarian Information Sheet - Alcohol*

12 Research

Colin Spedding

12.1 Organic vs conventional agricultural research

Since the production of organically-grown foods is based on the use of livestock and crop plants – just as in conventional agriculture – it is often suggested that the same research should serve both equally well. At a fundamental level, this is clearly true: all agriculture has to be based on the application of knowledge about how plants and animals grow, but it is also necessary to take into account the environmental context in which they grow.

Here there are some important differences. The environment varies, with topography, altitude, climate and weather, soil type and many other features, for all kinds of agriculture, but the history of conventional agriculture has involved continual attempts to modify or even control the environment. These attempts have included housing, protected cropping and the use of major inputs of agrochemicals (to change soil fertility and to control pests, parasites and diseases). In extreme cases, for example, intensive poultry and mushroom growing, the environment is under almost total control and can be optimised to suit the plant or animal species used. Indeed, even the breeds and varieties developed to suit different environments become unnecessary and largely disappear. Thus the breeds of intensively farmed livestock have gone from commercial agriculture, and pigs are known by code numbers rather than breed names. Only grazing animals, such as sheep, that occupy a wide range of uncontrollable environments, are still represented by large numbers of breeds (for example, there are over 50 breeds of sheep in the UK).

Organic farming, insofar as it is primarily land-based, differs radically from conventional farming in the extent and nature of the changes brought about in the environment. As a result, different breeds of animal and varieties of crops may be appropriate. This sometimes alters the characteristic problems but, even where the problems are essentially the same, the solutions required are quite different.

The organic philosophy, of 'working *with* nature', embraces complexity, whereas conventional farming aims at simplifying systems. This is not merely a question of differences in application, requiring different kinds of applied research; it also gives rise to different questions for more basic research. If, for example, pesticides are sufficiently effective in intensive agriculture, questions about biological control do not arise. And similar considerations apply to how plant nutrients are supplied, involving not only the use of legumes for nitrogen fixation in organic systems but also the encouragement of soil fauna to recycle plant materials. The list of these kinds of difference is a long one, but three general propositions can be identified.

(1) Fundamental research may, by nature, be similar whatever the intended application but, since not all possible research can be undertaken, the questions to be addressed

at this level still have to be selected, and these may be different for application in organic systems.

(2)   It is usually fairly easy to conclude that there is a problem, more difficult to decide exactly what it is and very hard to identify the appropriate solution (there are always a great many possible solutions). A simple illustration of this point, for conventional livestock production, is disease. Most diseases could, of course, be totally eradicated from a farming system but this might be quite uneconomic. Thus the problem for a veterinarian is commonly not 'how to prevent or cure a disease' but to do so in such a way that the farmer's profit is increased (or at least he is not made bankrupt by the solution). All such requirements have to be part of the definition of the problem. In the case of organic farming, the solution has to be consistent with the principles of organic production. So the relevant *applied* research may be different.

(3)   A 'systems approach' (Spedding 1988; 1996) is necessary whatever the agricultural system, but the greater complexity of organic systems makes this more necessary but also more difficult. It should be self-evident that, in all cases, the change to the system, implied or required by the chosen solution to a problem, must result in an improvement to the *whole* system in order to increase the achievement of its accepted goals, and not merely the solving of the specific problem. No one, for example would apply a spray to grassland, however efficient it was at controlling a weed problem, that resulted in poisoning the animals that grazed it.

Unfortunately, the consequences are not nearly so obvious for most changes considered. In addition to altering the environment, conventional farming has greatly changed the animals and plants used. Again, organic farming also uses selectively bred plants and animals and, again, the differences between this and conventional agriculture relate to both the extent and the nature of the changes brought about. The 'extent' theme is illustrated by a reluctance of organic farmers to go as far as conventional in the search for higher yields (especially in milk yield of cows, growth rate in meat animals and egg output in poultry).

The 'nature' theme is well illustrated currently by the aversion on the part of organic farmers to all aspects of genetic modification in the modern sense (GMOs and their derivatives are defined as organisms or products derived from such organisms which have been produced by the process of recombinant DNA techniques). The reasons for this aversion include both food safety and environmental concerns. This particular issue has very far-reaching implications as to whether organic and GMO-using conventional systems can even coexist in the same area (even when this is quite large – e.g. many square miles), due to the great distances that pollen can be dispersed by wind and bees.

The foregoing has implications for (1) the problems in organic farming that require research, (2) the nature of the research that is needed, (3) the organisational structure of the required research activity and (4) the way in which such research can be funded.

12.2   Problems requiring research

The first set of implications can best be discussed under three headings, related to crops, animals and processing.

12.2.1 *Problems of crop production*

All crop production, except protected cropping (for example, glasshouse, plastic covered), is vulnerable to the weather. Measures to combat frost and drought are common to most kinds of crop production: wind protection (by shelter belts and so on) is practised for only a few crops (e.g. top fruit). In general, apart from choice of location, little can be done about solar radiation, rainfall or temperature. Problems with weeds, pests and diseases are also common to all crops but the remedies available to organic producers are much more limited than for conventional farmers.

Organic standards vary across the world, from very strict to non-existent, but those in force in the European Union (EEC, 1991) list the substances that may or may not be used if food produce is to be sold with the label 'organic' in the UK (other EU countries have equivalent terms, such as '*biologique, 'ecologique'*). Where there are common standards, there are considerable similarities in the nature of the problems, although the species and incidence of weeds, pests and disease-producing organisms may be quite different in different countries.

The questions for research workers may thus vary widely but, in principle, will be quite different for organically-grown crops. Most of the chemical sprays are not permitted, so other remedies have to be found. This may mean that the problems for applied research are different and will often relate to whole husbandry systems and not just to specific practices. However, this is not confined to *applied* research. Interactions between plants and other organisms may need to be understood at a quite fundamental level, posing questions that simply would not arise in conventional crop production.

Weed control has to depend to a greater extent on cultivation methods and a greater understanding of weed species is needed than in circumstances where they can simply be sprayed out of existence. Pests and diseases may also be affected by cultivation methods but interactions with other organisms are of enormous importance.

The ecology of organic systems is generally more complex, and ecological research on natural (i.e. non-agricultural) systems has more relevance. This need to consider whole systems (often called a holistic approach) also includes what happens underground. The root system is commonly ignored, as if plants started at the soil surface. This is evident in the way most people view their lawns, recognising that the above-ground part grows and senesces (or is cut off) but not realising that roots are also growing and dying in a similar fashion. All this becomes extremely relevant in organic systems, since the organic philosophy is to feed the (living) soil rather than the plant directly (for example, with soluble nitrate fertiliser).

Similar considerations apply to animal production systems.

12.2.2 *Problems of animal production*

Although stockless systems are being explored, most organic farming systems require outdoor livestock as a component of the rotation. Experiments that investigate only one part of the rotation are therefore of limited value. Figures for output per hectare of cereal crops, for example are not easily comparable with those of conventional farming, whether expressed just for the cereal phase or over the whole rotation.

So the outputs of the livestock phase cannot be measured solely by their performance in terms of products but have to take into account their other contributions to the whole system. Unfortunately, there is no completely satisfactory way of comparing, for example, wheat with meat, or indeed combining them into one output figure. That is why ecologists tend to use energy and economists use monetary values, but the first does not take into account the different values put upon different types of energy and the second is subject to variation in prices. Economic assessments are, of course, required but they contribute little to identifying ways of improving the biological processes involved. Even economics has to accept that there are interactions and feedback in complex systems.

For example, feed costs are usually a very high proportion of the total costs of animal production, but no one would suggest that we stop feeding our animals in order to increase profit. The fact is that, whatever change we make in a system, it has multiple and often quite complicated effects on other parts of the system.

Thus the prevention and control of disease, for example, has to take account of this complexity, and simple remedies may have complicated consequences. Since organic animal production has to operate without the routine use of drugs, disease control has to depend on management, appropriate stocking rates, encouragement of natural resistance and appropriate use of animal species and strains.

Specific research needs will vary with the species, the system and the part of the world in which it is being operated, but solutions have to be found to control disease without the remedies available to conventional livestock production and, very importantly, without giving rise to animal suffering. Veterinary research is therefore needed that understands the constraints and objectives that govern organic animal production. In many cases, disease incidence is no higher and is sometimes lower than in conventional farming, but when it occurs there are fewer tested allowable remedies. 'Alternative' forms of medicine are easily dismissed because their effectiveness has not been scientifically proved but, in many parts of the world, people rely on naturally-occurring substances that have been found to work (often in ways that are not yet understood), for human medical treatment. Research is needed to explore the potential for such remedies in organic animal production.

12.2.3 *Processing*

Few agricultural products are eaten in their natural state and most are cooked. Increasingly foods are prepared by processing of crop and animal raw materials and there is little point in organic forms of production if the same principles are not applied during processing. Processing is undertaken for many reasons: the earliest was probably for preservation and storage, since many foods were highly seasonal.

Processing is now linked with ensuring food safety, especially to limit the growth of micro-organisms that may cause food poisoning as well as deterioration and wastage of food. Low temperature, acidification, heat and reduction of water availability (by adding salt or sugar, for example) have all been used for many years. But increasingly other methods have been developed, such as irradiation, vacuum packing and very much lower temperatures.

The problems for organic food production are that (a) the additives permitted for preservation (and to improve taste and texture) are greatly limited and (b) the substances used for cleaning processing machinery are also restricted. The research needs are therefore similar to those in production: how to achieve the desired aims with acceptable substances. Especially in developing countries, however, there are also problems of pest control during storage of the harvested products. Losses in storage and transport can be enormous and ways have to be found to reduce them substantially.

The UK Register of Organic Food Standards (UKROFS) regularly produces a list of specific needs for research in all these areas, but they naturally relate primarily to the UK (UKROFS 1995). Other countries have other specific needs.

12.3 Nature of the research required

It will have become clear from the preceding sections that a 'systems approach' is advocated as an essential feature of organic research. That is not to say that 'reductionist' experiments are never required. Quite often, in order to establish a causal connection between an apparent cause, and a recognised phenomenon (the postulated effect), it is necessary to design and conduct an experiment in which only one feature is varied within a much simplified system.

Two points have to be borne in mind, however. Firstly, the results of such experiments have to be applied within much more complex systems, and therefore have to be tested within them. Secondly, it does not always follow that because A causes B when only A is varied, that it will do so in the presence of factors C, D, and so on. Other factors may modify the effect of changes in A, in either direction. This is well illustrated in, for example, balanced diets (whether for plants, animals or people): increasing one dietary component may have no effect at all unless others are also increased. And this could be demonstrated for each component in turn; increasing each would have no effect. But, of course, increasing all of them simultaneously, in the right proportions, could have quite dramatic effects. It is this awareness, that the significance of a component can only be assessed in relation to the system as a whole, that characterises a systems approach.

Even where altering one component *does* have the right effect and solves the initial problem, it may well have many other effects, some of which may take a long time to show up. As has been remarked before (Spedding 1996), if you want to know what your next problem will be, have a look at your solution to the current one. So there is also a timescale requirement for organic research. After all, the minimum conversion period, before registration of an organic holding, is two years and in some cases it takes several more years before all the typical features (especially of soil fauna and flora) are established.

During the period of conversion, there may be special problems, applying methods of farming to a soil that is not yet entirely suitable. These problems are exacerbated if the *farmer* is also undergoing a conversion at the same time. Since fully organic products are not produced during conversion, the output does not attract premium prices: there are thus economic problems – although many countries now give grants to cover the conversion period. All this bears on the extremely difficult matter of experimentation on organic systems. Such systems evolve and take a long time to establish: they are quite unlike conventional systems, even conventional systems that produce a similar product.

Because of the complexity of organic systems, they are each more likely to be almost unique, presenting even greater problems in making comparisons and generalising from a result on one farm to others. This has always been the case for site-specific livestock systems (for example, grazing systems) compared with, say, intensive poultry or pig systems, which may be very similar wherever they are.

It is therefore extremely difficult to carry out relevant research without the involvement of those who understand organic principles. In the UK, there are a few research centres that are solely devoted to organic research (see Table 12.1): the best known (and longest established) are the Elm Farm Research Centre (EFRC) and the Henry Doubleday Research Association (HDRA).

The problem for applied research is how to combine relevance to the real world with scientific rigour and control. Objectivity is even more important where 'values' are deeply and necessarily embedded.

12.4 Organisational structures and funding

These two topics are inextricably linked. Some 10 years ago, it became fashionable in the UK to argue that 'near-market' research should be funded by the users. The stupidity of such a generalisation is well illustrated by medical research. Could a cancer-sufferer be expected to fund the research he did not know he was going to need or is the research cost to be added on to that of treatment at a time when he is least able to pay? The weaknesses in the generalisation are such as to render further illustration superfluous.

Every research area could provide examples of the damage done by the blanket application of the proposition. Even where it seems most relevant, there are regrettable consequences. If a drug company has to fund even the most fundamental research that might lead to profit, it will only embark on the development phase where a return on the huge investment can be foreseen. Thus minority requirements are of no interest: small markets (for example, minor crops, diseases that only affect a few people or only poor people), air or sea pollution where no owner can be identified, and research where there is no product or service to sell – all will be ignored.

The fact that the whole idea was flawed, based as it was on a grossly over-simplified model of the way research works, has been pointed out repeatedly. The Institute of Biology, for example, produced a Report (NRPG 1991) that illustrated the inadequacy of the linear model (fundamental → basic → applied research → development → markets), pointing

Table 12.1 UK organisations involved in organic research.*

Organisation
ADAS – Pwllpeiran
ADAS – Redesdale
ADAS – Terrington
Centre for Organic Agriculture, Aberdeen
Elm Farm Research Centre, Newbury
Greenmont College of Agriculture, Northern Ireland
Henry Doubleday Research Association, Coventry
Institute of Grassland & Environmental Research, Aberystwyth
Scottish Agricultural College Organic Farming Unit, Aberdeen
Veterinary Epidemiology & Economics Research Unit, Reading
Welsh Institute of Rural Studies Organic Farming Unit, Aberystwyth

*Further details are given in Lampkin & Measures 1999

out that research is full of feedback loops and can originate at any point in the chain, and described the likely consequences of the (government) policy (see Table 12.2).

Those consequences are now with us. Universities have been obliged to seek funds from industry, university researchers have been required to obtain money from industry, and commercial firms conduct their own vast programmes. The result is that it is now quite difficult to find wholly independent scientists: many of the most expert scientists in any field are funded by an interested party in that field. Where does one then go for independent advice? Even freedom to publish may be constrained or delayed, to preserve commercial advantage for the funder.

This is a subject of immense importance but here we have to concentrate on the consequences for organic research. Clearly, the small size of the organic sector, the fact that it minimises the purchase of manufactured inputs and that it avoids unnecessary processing, all represent a relatively negligible opportunity for commercial exploitation by support industries.

Perhaps only the retailers have a commercial interest in the improvement of efficiency in organic farming, for two main reasons. Firstly, the integrity of the production process and the continuation of that integrity to the point of sale, is vital to the credibility on which the organic market depends. This covers many of the research needs of organic farming. Secondly, the demand for organic food products is increasing at a remarkable rate, throughout Europe and beyond. Presumably, the market would be even greater if price premiums were lower. If that is a proposition of interest to retailers, then it would be worth considering whether research could improve the economic efficiency of organic farming to such an extent that these premiums did not need to be so high. Insofar as they reflect

Table 12.2   Consequences of leaving R&D to be funded through the market.

No.	Possible outcome
(1)	Certain sectors of activity might not be supported by R&D at all
(2)	Areas that could not generate large profits would be neglected. (Even if the activity was extensive, it might be difficult to achieve a profit or, indeed, any return at all to those who fund the research.) Alternatively, even if the expectation of profit were sufficiently great in a mathematical sense, the very high risk associated with this might be too much for individual commercial organisations to accept
(3)	Some important areas might not be served at all, because the information generated might be counter-productive to the commercial interests of those most involved in the commercial activity
(4)	Results of some R&D might be revealed selectively: only those results favourable to commercial interests might be published and those unfavourable might be suppressed
(5)	The information publicly available might not always, therefore, be regarded as credible or reliable – and certainly it could not be regarded as independent. In some cases, this would not be of serious concern and, in the long term, would affect commercial reputation and success
(6)	There might be very little contribution to the 'community of science' and to its publicly available, peer-reviewed literature. This is important because ultimately most scientific advance has to be based on the accumulated knowledge produced within this framework
(7)	Unnecessary duplication might result from the fact that some research remained unpublished, for reasons of commercial secrecy
(8)	Attention would not be drawn to side-effects, unforeseen consequences, effects on welfare and the environment, associated with the products sought or the production processes used
(9)	To the extent that publicly-funded scientists were mainly engaged in research divorced from obvious application, the public perception of scientists as occupants of 'ivory towers' might be reinforced

a current shortfall in supply relative to demand, they may not survive at these levels anyway. It may be necessary, therefore, to learn how to lower the costs of production: a major route to this would be to increase yields per unit of resource (not *necessarily* per man, per animal or per hectare).

This has important implications. For example, if organic farming could greatly increase herbage yields per hectare, would the resultant increases in the stocking rates of grazing animals cause problems with internal parasites? It is worth noting that this problem (of parasitic infestation and stocking rate) was being studied in conventional grassland farming over 40 years ago. In spite of the superficial linkage, the burden of parasites is related to the animal's intake of infective larvae and thus the number of larvae per kilogram of herbage. And since the amount of herbage has to be in proportion to the number of animals carried, there is no necessary increase in infestation, simply as a result of increased stocking rate.

This is an example of a practical problem of considerable importance, where the relevant research would certainly be described as 'applied'. Yet it would require quite specialised inputs of scientific information (for example, the life-cycles of parasitic nematodes) to be harnessed within a systems context. We are thus brought back to the need for a systems approach to research and the questions here is: how can that be organised and funded?

There is no doubt that systems research requires a multidisciplinary team, in which all relevant specialisms are represented – not by those who are second-rate, but by those who have a high standing in their subject and are thus able to harness their branch of science to the solution of multidisciplinary problems. Furthermore, these specialists need to continue in their own fields, in order that they should be both up-to-date and credible representatives of it.

Most successful leaders of systems research, in whatever country, have emerged after achieving distinction in their own specialisms because they wish to solve problems of practical importance and recognise that their discipline has a contribution to make. It is a small step for them to recognise that this also applies to many other disciplines as well.

Such teams require leaders with a grasp of how all these specialisms fit together: indeed, their leadership depends upon them having a better picture of the whole system than the other members of the team. They also have to have an ability to communicate with the others, cutting through any jargon. High-quality specialists are more likely to possess this facility, simply because only those who really understand a subject are capable of expressing their knowledge in basic terms, yet without oversimplification. Team leaders are sometimes dismissively referred to as 'generalists'; in fact, they are more akin to 'generals'.

Who is going to develop such teams and who is going to fund them? This needs to take place against a background in which advancement in science is seen as being more likely to flow from increasingly narrow specialisation. This has been evident with those who build the mathematical models used in systems research (mathematical because this is necessary to cope with the sheer complexity of the systems modelled). Very often, the modellers become specialists in mathematical modelling techniques and cease to be involved with the systems themselves.

Apart from interested parties (such as the retailers already mentioned), there are disinterested parties, notably charitable foundations, research councils and government.

In the UK, all are involved and MAFF is becoming so to an increasing extent, with a 40% increase in funding for organic research for the year 1999–2000.

A recent research study into the environmental implications of organic farming was supported jointly by the Biotechnology and Biological Sciences Research Council (BBSRC), the Economics and Social Research Council (ESRC) and the Natural Environment Research Council (NERC), and was carried out by the Institute of Arable Crops Research (IACR), the Institute of Grassland and Ecological Research (IGER), the Wildlife Conservation Research Unit (Wild CRU), the Institute of Terrestrial Ecology (ITE), and the School of Environmental Sciences at the University of East Anglia (Cobb *et al.* 1998).

12.4.1 *Government funding*

Whoever funds organic research in a substantial way still has to assess priorities – since not everything can be supported and there has to be accountability for the expenditure of funds, whether public or private. It is therefore necessary to have a national focal point, at which a whole picture of the research effort can be assembled, with up-to-date information on who is doing what and who proposes to do what, as well as what has been done and published in the public domain. Against that background, priorities can be assessed, with reference to existing and past research, but only by those able to assess the relative importance of the topics proposed and the likelihood that the research proposed will lead to useful advances (in reasonable time and at an acceptable cost). How is this to be done?

It is best done by establishing a group that reflects all the necessary skills to make the required judgements and to work together. It is more difficult if it is based on a point at which only some of these judgements can be made and which then consults more widely. In terms of structures, it is only possible to suggest mechanisms for specific countries.

In the UK, substantial public funding of research is channelled through research councils. Indeed, industry funds may also be funnelled through research councils. Thus, the BBSRC (Biotechnology and Biological Sciences Research Council) and the other major research councils are publicly funded, and organic production may fall within their remit.

But the Apple and Pear Research Council (APRC) is entirely funded by a compulsory (area-based) levy on growers, and the Horticulture Development Council (HDC) is similarly levy-funded with the levy based on a percentage of turnover. There are also other sector-based levy bodies, some without a statutory function and thus only empowered to impose a voluntary levy on producers. Hence research bodies use a variety of funding methods and a particular model does not have to be followed. Indeed, it is probably better to start out with no preconceived ideas but to consider what would best serve the needs of the sector. No serious thought has yet been given to establishing a research council for the organic sector, although the idea has been suggested a number of times.

UKROFS does, at the request of Ministers, compile a list of R&D needs (UKROFS 1995; 1999), based on considerable consultation, but the UKROFS board has no resources to fund research and its efforts have been sporadic rather than continuous. The list is substantial and only the highest priority topics can be illustrated here (Table 12.3).

Identifying gaps and assessing priorities is useful but UKROFS can only recommend that these gaps should be filled and has no role in deciding how or where this should be done or in monitoring the progress of the resulting work. Current MAFF projects are shown in Table 12.4. Further, there is no obligation on any research organisation to keep UKROFS informed as to current or proposed R&D in organics.

There is much to be said for mixed funding, so that money can be accepted for additional work in specific areas, but an Organic Research Council would have to be independent in its operations, in commissioning and monitoring the research programme. Publication of results through reputable and available channels would have to be ensured. It would probably be essential to have a publicly funded administration and a core levy to establish an independent and balanced research programme. It is highly desirable that there should be a statutory basis for a compulsory levy (on beneficiaries of the research) and levy payers would sensibly be represented on the Council. However, there should also be powerful independent representation and the chairman should be completely independent.

There is no reason why ideas of this kind should not be freely debated by all concerned before taking any firm steps and without prior commitment. The motivation should be to determine what is needed, in the interests of an orderly development of the sector. There would almost certainly be considerable spin-off for conventional farming and it would be sensible to ensure adequate liaison with other relevant funding bodies.

Table 12.3 Examples of top priorities for UK organic R&D (source: UKROFS 1995; 1999).

No.	Topic
(1)	Develop viable weed control strategies for agricultural and horticultural crops
(2)	Develop plant propagation systems for modular transplants and vegetative reproductive material (e.g. potato tubers, onion sets)
(3)	Methods of control of external and internal parasites of sheep, cattle and poultry using approved substances
(4)	Fate and transfer of nutrients between organic manures and plants
(5)	Pest and disease control in horticultural crops
(6)	Control of seed-borne and seedling diseases
(7)	Identify problems of conversion on different farm types
(8)	Economics of stockless systems and criteria for organic poultry production

Table 12.4 MAFF funded horticultural R&D topics.

Status	Project
Current projects	Variety trials
	Transplant production
	Herbage legume intercropping
	Organic horticulture conversion
	Technical and economic problems of protected crop and apple production
	Storage of organic crops
	Review of organic fruit production
	Economics of organic fruit production
New projects tendered	Disease control strategies for organic vegetable crops
October 1998	Potato blight control
	Organic vegetable seed production
	Companion cropping for organic field vegetables

Finally, there is a need to consider whether the organic sector (in the UK) should be limited by the boundaries set by Regulation EC2092/91. This only covers foodstuffs and unprocessed non-food crop products and, while there does not seem to be much sense in including such products as 'organic' shampoo and the like, it seems illogical not to include all agricultural products, such as wool and fibrous crops, oilseeds, fuel crops, by-products of both crops and livestock (for example, leather, bonemeal) and plants that produce medicines, perfumes and flavourings, even when they are processed, provided that the processing also meets organic criteria. After all, the principles of organic production do not just relate to food qualities (safety, taste, and so on) but also to the effects on the soil and environmental impact.

12.5 Environmental impact

If organic farming is to preserve or improve the environment, there is a need for research into ways of doing so. Environmental impact has to include effects on landscape, biodiversity and the preservation or encouragement of particular species of fauna and flora. In some cases, the relatively small areas devoted to organic farming might not be adequate to affect wide-ranging species, notably larger predators whose territories greatly exceed the areas of many farms. If this is so, then there may be additional arguments in favour of large-scale organic farming or the aggregation of units into compact areas. This also has implications for such problems as spray drift and the spread of GM pollen.

Similar arguments can be developed for research into animal welfare, human health, rural employment, and contributions to the rural economy. The most urgent need for research in these areas is to produce evidence to support the commonly-made assertions that organic farming is beneficial in all of them – assuming that there is, indeed, evidence to give such support (see MAFF 1995).

However, one should not get trapped into sweeping generalisations about either organic or conventional farming. There is enormous variation in both, in species and varieties used, level of intensity, scale, soil type, climate and weather, productivity and profitability. This is true within a country but, of course, the variation between countries may be enormous and both research needs and institutions (see Table 12.5 for European examples) will differ greatly.

Table 12.5   Examples of European organic research organisation (source: IFOAM 1999).

Country	Organisation
Denmark	CENVIR – Centre for Ecology and Environment
France	GRAB – Groupe de Recherche en Agriculture Biologique
Germany	Institut für biologisch-dynamische Forschung
Greece	SOGE – Association of Organic Agriculture of Greece
Italy	AIAB – Associazione Italiana per l'Agricoltura Biologica
Netherlands	Agro Eco Consultancy
Norway	NORSØK – Norsk Senter for Økologisk Landbruk
Switzerland	FIBL – Forschungsinstitut für Biologischen Landbau

Therefore any generalised comparison of organic and conventional farming that does not take these differences into account cannot have any useful meaning or validity.

References

Cobb, R., Feber, R., Hopkins, A. and Stockdale, E. (1998) Organic Farming Study. *Global Environmental Change Programme Briefings*, No. 17, March, 1998.

EEC (1991) Council Regulation (EEC) No. 2092/91.

IFOAM (1999) Organic Agriculture Worldwide, IFOAM Directory of the member organizations and associates 1998/99.

Lampkin, N. and Measures, M. (1999) *1999 Organic Farm Management Handbook*, 3rd edn. WIRS & EFRC.

MAFF (1995) *The Effect of Organic Farming Systems on Aspects of the Environment*, A Review prepared for Agricultural Resources policy Division of MAFF.

NRPG (1991) *The Public Funding of Research and Development in Biology*, A Policy Study prepared by the Natural Resources Policy Group, IOB Policy Studies No. 2, June 1991, Institute of Biology, London.

Spedding, C.R.W. (1998) *An Introduction to Agricultural Systems*, 2nd edn. Elsevier Applied Science.

Spedding, Colin R.W. (1996) *Agriculture and the Citizen*. Chapman and Hall.

UKROFS (1995) List of R&D Priorities for Organic Production. UKROFS, London.

UKROFS (1999) List of R&D Priorities for Organic Production. UKROFS, London.

13 Australia and New Zealand

Scott Kinear

13.1 Australia

13.1.1 *Brief history of organic and biodynamic movement*

In discussing the beginnings of organic and biodynamic production systems in Australia, it is important to acknowledge those indigenous people who have practised sustainable farming for generations. A further group of twentieth-century farmers, like their counterparts in Europe, observed the damage caused by chemical intensive agriculture and chose to listen to the land in a way not unlike the indigenous peoples before them. Some of these people went on to formalise the organic and biodynamic movement with the establishment of the Biodynamic Research Institute in 1967 and then the Organic Retailers and Growers Association of Australia (ORGAA). These were followed by the National Association for Sustainable Agriculture Australia (NASAA), the Biological Farmers of Australia (BFA), and the Organic Herb Growers of Australia (OHGA) in 1986. A few years later the Organic Vignerons Association of Australia (OVAA) and the Tasmanian Organic-Dynamic Producers (TOP) was established. In more recent years the Organic Food Chain (OFC) formed.

In May 1998 the peak industry body, the Organic Federation of Australia, was formed to promote and lobby on behalf of the organic industry. In addition it aims to assist in resolving industry issues best addressed with a national whole of industry focus. Australia also has a certified organic retailer scheme managed by the Organic Retailers and Growers Association of Australia established in 1990. This was implemented to promote certified organic and biodynamic produce traded in a guaranteed manner for consumer and retailer protection.

13.1.2 *The agricultural landscape in Australia*

While New Zealand can truly claim to be 'green', Australia also has extensive 'green' agricultural zones along the south and eastern seaboard and in the southwest of Western Australia. The centre of the country is without doubt arid semi-desert, though of course this is the basis of a significant tourist industry which is now one of the largest earners of export income for Australia.

Despite having this 'green' strip along the coast, Australian soils are ancient and fragile. Much of Australia was covered by inland seas millions of years ago, which have left a high salt load in the subsoils. Jason Alexandra (1999) states:

'The environmental challenges facing Australia's land and fresh-water based industries are huge and widely recognised. The stakes are high in both economic and ecological terms with numerous degradation trends, such as dryland salinity, loss of biodiversity and declining water quality well established. Ecosystems suspected to be at risk include national treasures such as the Murray Darling Basin [the Murray Darling River system is responsible for supplying 80% of Australia's irrigated agriculture], many other rivers and estuaries, the Western Australian wheat belt and the Great Barrier Reef.'

The last 150 years of farming have wreaked much havoc with lowered soil fertility and soil erosion. Australia's overuse of water resources for irrigation, which still continues, despite dire predictions from leading soil and water scientists (Harris 1999), has caused an enormous area of land to be salt-effected thereby increasing the salt load in major river systems. Land clearing has caused dryland salinity where water rises to the surface bringing salt that would have remained if the tree cover had not been removed. Successful agriculture can exist on much of the fragile agricultural zones of Australia but only if clearing of the native cover is limited. A balance needs to be struck between cleared and timbered cover on each property, with the regional management of irrigation water an essential complementary strategy.

There is a growing awareness of water use issues, with a recent state government election result influenced by promises of returning an environmental water flow to the Snowy River in Victoria. Such a move would see a decrease in water flowing down the Murray Darling River system, thereby reducing water available to farming communities in South Australia. Consideration of radical options such as these is absolutely essential if Australia is to achieve sustainable use of water and land resources. If this issue is not addressed some commentators are predicting that Australia will become a net importer of food in just 50 years.

13.1.3 *Market overview of production*

While its natural resources are fragile and degraded Australia is still well positioned to produce genuinely clean organic and biodynamic foods. Compared to Europe and North America, farming in Australia has generally been carried out at much lower intensity and soils and rainfall are virtually unaffected by pollution. In addition, Australia takes full advantage of the range of growing seasons down the east coast from tropical in the north, to temperate in the south, producing a wide range of organic and biodynamic fresh foods of the highest quality.

The organic industry in Australia has now grown to number approximately 1600 certified producers (Table 13.1). The total area of land certified is in excess of 7m ha with large tracts of this area comprising rangeland cattle country in outback New South Wales and Queensland.

Figure 13.1 shows the projected growth of the organic industry in the most recent report on its status by Macarthur Agribusiness, & Quarantine and Inspection Resources Pty Ltd (1999). These growth rates are conservative given the increase in domestic demand in the last 5 years that has exceeded 20% growth per year (Williams 1999). Some of this demand is being met by import of processed organic products not manufactured in Australia, such as breakfast cereals and chocolate.

Table 13.1 The number of certified growers and processors in Australia and area of land for each certifier.

Organisation	Certified growers and processors	Approximate area (000 hectares)
BFA	544*	150*
BDRI	144†	not available
NASAA	483‡	7500‡
OFC	18*	not available
OVAA	20*	0.12356*
OHGA	436§	4*
TOP	25¶	0.8¶
Total	1670	7655

Sources: *McCoy & Parlevliet (1999); †Macarthur Agribusiness, & Quarantine and Inspection Resources Pty Ltd (1999); ‡NASAA advice, current as at 30.6.99; §OHGA advice, current as at 9.12.99; ¶TOP advice, current as at 28.1.99.
BDRI = Biodynamic Research Institute
BFA = Biological Farmers of Australia
NASAA = National Association for Sustainable Agriculture Australia
OFC = Organic Food Chain
OHGA = Organic Herb Growers of Australia
OVAA = Organic Vignerons Association of Australia
TOP = Tasmanian Organic-Dynamic Producers

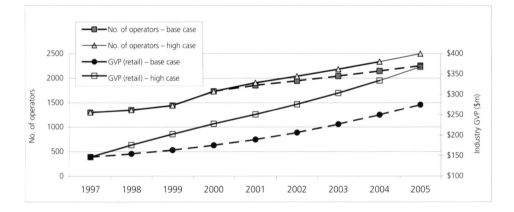

Fig. 13.1 Forecast growth: number of operators and gross value of product (farm-gate) (source: Macarthur Agribusiness, & Quarantine and Inspection Resources Pty Ltd 1999).

13.1.4 *Organic industry organisations*

Organic industry peak body

The Organic Federation of Australia (OFA) was established in 1998 to represent the interests of the industry to government, the media and the public. It campaigns heavily on the issues affecting the organic industry such as domestic regulation of organic labelling and protection of organic from genetically engineered crop contamination. In addition the OFA continues to lobby for increased government support for organic research and development and advisory, education and extension services.

Certification organisations

There are currently seven independent grower-run certification bodies which are all accredited by the Australian Quarantine Inspection Service (AQIS) to meet export standards set by the European Union.

The National Association of Sustainable Agriculture of Australia (NASAA) began in 1986 when its production standards were first published. Since 1994 NASAA has been accredited by the International Federation of Agricultural Movements (IFOAM), as well as AQIS, and remains the only certifier in Australia to date to achieve IFOAM accreditation. NASAA has played a pivotal role internationally with its position in the international organic arena well recognised. NASAA has assisted in representing Australia at numerous Codex Alimentarius organic committee meetings. NASAA certifies 2000 producers and processors overseas in Papua New Guinea, Sri Lanka, Indonesia, Japan and Nepal. In addition NASAA is a signatory to the 1999 multilateral agreement between IFOAM accredited bodies facilitating transference of certification between such operators internationally. While NASAA has predominantly certified organic production systems, it has recently begun to certify biodynamic farmers.

The Biological Farmers of Australia (BFA) was established in 1988 and now comprises a broad-based membership of 1200 including farmers (49%), processors (9%), wholesalers, retailers and exporters (3%), input suppliers (3%), and associate members (36%). The BFA has also begun a system of certification of retail outlets similar to the Organic Retailers and Growers Association of Australia scheme (see section below on retailer certification). The BFA certifies both organic and biodynamic production processes. Itself a product of a breakaway from earlier organisations, the BFA has traditionally been farmer-operated with a strong focus on a mentor system of organic operators.

Alex Podolinsky established the Biodynamic Research Institute (BDRI) in 1967 and he still works actively inspecting and advising biodynamic farmers for the institute. The BDRI is the holder of the Demeter logo in Australia for biodynamic produce, with significant certifications in broadacre (wheat), livestock (beef) and horticulture. So far it is the exclusive certifier of biodynamic product carrying the Demeter logo to Europe, however a different organisation, the Biodynamic Farmers and Gardeners Association of Australia (BDFGAA) is attempting to get Demeter recognition for its products sold into Europe. (The BDFGAA is not a certification organisation though various options are under consideration. At present its members number approximately 400, both certified and uncertified, with the certified members mostly accredited by the BFA and NASAA.) (See section 13.1.5.)

The Organic Herb Growers of Australia (OHGA) was established in 1986 to foster the development of organic herb growing in Australia. In the early years it began certification focused on herb growing and processing, though more recently it has been certifying a diverse range of organic enterprises. It offers one of the lowest cost certification schemes in Australia.

The Organic Vignerons Association of Australia (OVAA) is a small certification organisation dedicated to organic winemaking. Its membership is small yet some of its members are very successfully exporting product into Europe and the US.

The Organic Food Chain (OFC) was established only a few years ago because of dissatisfaction with some of the other certification bodies. A number of OFC members are extensive grain growers in Queensland, involved in large export sales. The OFC holds ISO 9002 as well as AQIS accreditation and aims to assist with marketing and networking.

The Tasmanian Organic-Dynamic Producers Association (TOP) was established in the early 1990s, operates entirely within Tasmania, and is the most recent organisation to receive AQIS accreditation. Tasmania as an island state south of mainland Australia, has good rainfall and excellent conditions for horticulture, dairy and beef production. The organic industry is highly active in Tasmania, working to keep the state free of genetically engineered organisms and on track to go totally organic by 2020 (a vision articulated by organic and environment groups in 1999).

One problem that will need to be overcome is a significant forest industry which poses a threat to organic from the overspray of pesticides and water runoff. This will be a major issue for TOP and the OFA to address in 2000.

Retailer certification

The Organic Retailers and Growers Association of Australia (ORGAA) has been in existence since 1986, beginning with a strong farmer base. In the early 1990s it began a system of retail certification which was a world first, similar to accreditation of farming systems. There are now more than 50 specialty retailers throughout the country that are certified with ORGAA. Through a national freecall number consumers can telephone to find out their nearest certified outlet. With the absence of Australian domestic regulations to control the use of the words 'organic' and 'biodynamic', this system has provided a welcome guarantee to consumers. While all the large fruit and vegetable wholesalers are trading exclusively certified organic produce, only approximately 20% of the organic retailers around Australia are signatories to the ORGAA certification system. Thus opportunities for fraud at this level are numerous and hard to police.

In addition to ORGAA, the Biological Farmers of Australia, as previously mentioned, also certify a small number of retailers around the country.

Advisory and extension in Australia

There are a number of industry groups offering advice and assistance to growers though there is a need to establish more regional grower groups. ORGAA provides the Organic Advisory Service which for the last 10 years has provided information about organic foods to students, farmers, government departments and industry. The Biodynamic Farmers and Gardeners Association of Australia has also been pivotal in establishing regional groups of biodynamic farmers. It offers educational courses, now with government accreditation, which allow farmers to attend with some of their costs reimbursed by government. This arrangement was announced at the beginning of 2000 and is seen as a significant step forward in the organic industry quest for recognition by government. It is important to mention the Canberra Organic Growers Association based in Australia's capital city. It has been actively promoting Australian organic food on one of the first and most informative organic websites set up in this country.

While the certification organisations are a good point of contact, organise regular field days. and can refer people on, they do not provide a significant level of advice to farmers.

13.1.5 *Regulations for production and processing in Australia*

At the request of the Organic Retailers and Growers Association of Australia (ORGAA began as a Victorian organisation and then in 1993 grew to become a national organisation) and other industry associations (NASAA, BFA and BDRI) along with the Victorian Department of Agriculture in the early 1990s, the Organic Produce Advisory Committee was set up by the then Federal Primary Industries Minister. This committee was asked to develop a national standard and to advise the minister on matters related to organic foods. The committee consisted of a number of organic industry and government representatives, and a consumer and environment representative.

In 1992 the first national standard for organic and biodynamic food production was released and a system of accreditation by the Australian government was negotiated using the Australian Quarantine Inspection Service (AQIS). Ruth Lovisolo deserves special mention as the AQIS officer responsible for working with industry to implement the system of accreditation. She is also well known in international organic circles as the chair of the Codex Alimentarius Organic Committee, which has undertaken the mammoth task of developing world standards for production, processing and labelling of organic foods.

Negotiations with the European Union, after development of the national standard in 1992, conferred 'listed third country status' on all organic and biodynamic foods certified by organisations accredited by AQIS. (Listed third party status is conferred by the European Union on those countries that have demonstrated government-to-government equivalence to the EU organic standards.) AQIS acts as the competent authority and is responsible for the issuing of organic export certificates, which must accompany all exports labeled organic or biodynamic. In late 1999, approval was given for the individual listing of the AQIS accredited certification agencies who write their own export certificates.

Until 1998 there was no law that prohibited the export of organic foods that were uncertified. A voluntary code of practice existed which saw the bulk of organic foods, certainly those into Europe, carrying certification. In October of 1998 'organic export control orders' were implemented, making it illegal to export any food or fibre product from Australia labelled organic or biodynamic unless it was certified by one of the AQIS accredited certification agencies. This covers exports of organic or biodynamic products to any part of the world. Unfortunately Australia does not offer the same protection to domestic customers and this has been a point of contention between the organic industry and government regulatory authorities.

While the Australian national standard for the production of organic and biodynamic foods meets international requirements, especially those of the European Union, there is significant interest in developing regional standards, which recognise the unique growing conditions experienced in a country like Australia. Within Australia there are tropical, temperate, arid and semi-arid growing regions that are in some instances similar to Europe and in others vastly different. Liz Clay is Australia's regional IFOAM representative and she argues this case in point adding, 'there is good grounds for a regional approach to international organic standards developed by IFOAM'. She argues that IFOAM is best positioned to provide world standards but that it must recognise the particular regional differences that apply unique pressures to organic production systems. For example, many north European countries experience strong winters that assist in breaking disease cycles whereas, in Australia, milder winters mean the same diseases can impact significantly on production.

13.1.6 *Market overview of exports*

Hassell & Associates, in 1996, reported exports at approximately A$30m. The Organic Federation of Australia estimates current exports are approximately 30% of production at A$60–80m. This export percentage is low considering the national average for food exports from Australia is 80%, indicating that considerable untapped opportunities exist for organic exports. After European settlement the economy of Australia was built on the export of agricultural commodities for many years. Products like wool and wheat were the major export earning commodities before mining and tourism took over in the latter part of last century. While some traditional export markets may be decreasing (due to the importing country achieving a higher level of self-sufficiency) there are still significant markets in Asia, such as Japan, who are dependent on importing food to sustain a food supply.

Organic exports go to a range of countries and most are commodity based (Table 13.2). With a long history of supplying commodities, successful attempts by Australian companies to break into the processed organic overseas food markets are few. The further development of the export market is seen as essential if Australia is to develop a robust organic industry. The OFA plans to form an organic export advisory group to facilitate export development similar to the Organic Products Exporters Group (OPEG) in New Zealand, which has been highly successful. Key opportunities exist in local Asian markets such as Japan, Hong Kong, Singapore, Malaysia, Thailand and Indonesia. A recent Victorian partnership has seen Japanese investment in a noodle factory which is now very successfully exporting 2000 tonnes per year of organic noodles to Japan.

Many inquiries (often from Japan) have been received in the last few years that far exceed product supply capability. The US, and now some European countries, are increasingly geared to export of organic and there is significant competition for the Asian market. To succeed in Australia, long-term contracts, well in advance of supply, are essential with overseas buyers. This will encourage investment in conversion and production infrastructure, sufficient to supply large quantities of organic product to overseas markets. It is important that Australia continues to highlight the comparative advantages of government accreditation based on a strong national standard and a relatively unpolluted environment.

13.1.7 *Market review of domestic consumption*

There is little data available to profile the organic industry in Australia and anecdotal evidence is used to extrapolate from the most recent survey by Hassell & Associates 1996 who reported total sales of A$80.5m. (see Table 13.3.) The two largest wholesale outlets for organic fresh fruit and vegetables in Melbourne and Sydney have reported consistent growth over the past 5 years of 25% per annum. The same Melbourne wholesaler recently reported 60% growth in the last 6 months, due most likely to increased media coverage

Table 13.2 Current Australian organic and biodynamic exports (source: McCoy & Parlevliet 1999).

Product	Destination
Apples	Europe, Germany, Holland, UK
Bananas	Japan
Barley	Switzerland, Japan
Beef	Japan
Biscuit mix	Japan
Canola oil	Japan, US
Carrots	Malaysia, Singapore
Chickpeas	Holland
Essential oils	Europe
Eucalyptus oil	Europe, US
Flax oil	Japan, Hong Kong, Malaysia
Flour	Japan, Italy
Flour pre-mix	Japan
Honey	Singapore
Juice – fruit or vegetable	Japan
Juice – orange	Japan
Linseed	Holland
Malt – beverage	Japan
Mayonnaise	Europe, US, Japan
Mixed fruit and vegetables	Singapore, Malaysia, Hong Kong
Mung beans	Holland, Germany, US, Italy
Oats	Switzerland, Japan
Orange – Valencia	Sweden
Oranges – navel	Europe, US, Holland, Germany, UK
Pears	Germany, Holland
Processed products	Southeast Asia, Japan, Europe
Rice	Switzerland, Malaysia, Japan, Europe, US
Safflower oil	Japan, Germany, Holland, Switzerland, France
Salad dressing	Europe, US, Japan
Soybeans	Japan, Holland, Germany, Europe
Sunflower oil	Japan, Holland
Triticale	Japan
Wheat – durum	Italy
Wheat – hard	Austria, Switzerland, Japan, UK, Holland, Norway, Sweden
Wheat – noodle	Japan
Wine	Europe, UK, Japan
Wool	Germany, Japan

of organic food because of consumer concerns related to genetically engineered foods. Current domestic sales are estimated by the OFA to be in the range A\$150–200m. This is 1–2% of production and consumption in Australia.

Most organic foods are sold in specialty stores, either dedicated organic stores or in health food and natural foods stores. Demand is greater than supply for a small but significant percentage of food lines at any one time, though definitely not for the majority. The quality of foods has increased significantly over time with a resulting increase in consumer interest and attraction of gourmet buyers searching out the freshest and tastiest products.

Supermarkets, although they have dabbled in organic foods over the last 10 years, are now carefully planning their sales strategy. Woolworths, a national supermarket chain with more than 600 stores, has fresh organic foods in 57 stores as reported at a conference in Mackay, Queensland, in September 1999. Coles, the other major retail chain, has set up national certified prepacking and processing operations for fresh organic foods. The main concern of supermarkets is the quality and consistency of supply. The future should see long-term contracts offered in an attempt to stimulate production quality and quantity consistent with mainstream retailing of organic foods. One problem with supply is that all the growers of a particular product at any time of the year can be in the same topographical area, and uncontrollable forces can have a devastating effect. As the industry gains expertise and expands further this effect will be minimised.

13.1.8 *Organic industry profiles*

Processors

Flour and grain. Companies growing and then processing organic flour and grain were the first to develop in Australia, some more than 30 years ago. This led to the establishment of distribution companies in capital cities to supply the health food and bakery industry. With a small population base of around 18 million, Australia has relied heavily on imported organic processed products. Low domestic consumption has not often justified the investment in machinery and segregation arrangements required to process many organic foods. That is steadily changing with investments in processing plant and machinery in the million dollar range and demarcation within large plants, though the majority of products are still from small regional plants and factories. There is a need to gain further access to export markets to justify serious dedicated infrastructure development for organic processing. This is often difficult, given the competition from European and US organic processors who are well developed with huge markets on their doorsteps. Australian companies need to focus on producing unique products, which will create lasting export potential in overseas markets. One example is a growing interest in the organic certification of Australian bush foods. This has the potential to capture unique international interest. In the last 10 years processors have developed a wide range of products from dairy, bread and biscuits to jams, pasta and wine.

Four Leaf in South Australia began selling organic produce in 1968 and became certified in 1988, producing a wide range of stoneground flours and cold pressed oils. The company farms 3000 acres with a mixed enterprise property, also producing organic lamb and wool, which has been sold for 'eco' sweaters in Europe. The oils have been sold into New Zealand the US and Canada, and some grain products to Asia and South Africa.

Kialla Pure Foods is Australia's largest certified organic grain processor, located in Queensland two hours west of Brisbane. Like Four Leaf, Kialla was first a farm and then a processor, though now it buys from other growers to supply the domestic and export markets with a large range of flour and grain products. Its export clients are located in New Zealand, Malaysia, Taiwan, Hong Kong, Japan, Switzerland and the UK. Grains are milled with traditional solid millstones and are stored in silos sealed with carbon dioxide or in controlled atmosphere cold rooms. Kialla has recently added a $2500 \, m^2$ factory to its operation, making this the most modern dedicated organic grain and flour processing plant in Australia.

Table 13.3 Breakdown of Australian organic industry retail sales, 1995 (A$m) (source: Hassell & Associates 1996).

Product grouping	Queensland	New South Wales	ACT (Canberra)	Victoria	Tasmania	South Australia	Western Australia	Total
Livestock products	0.3	5.3	0.06	1.2	0.02	0.2	0.4	7.48
Seeds, grains, cereals	3.2	1.4	0.36	3.1	0.28	0.9	1.2	10.44
Fruits, nuts	4.9	7.9	0.39	11.4	0.35	1.3	1.2	27.44
Vegetables, herbs	6.1	6.7	0.32	11.4	0.24	1.7	1.6	28.06
Tree products, natural oils	0	2.2	0	0.3	0.01	0	0	2.51
Other	0	0.5	0.07	3.4	0.10	0.3	0.2	4.57
Total organic	14.5	24.0	1.2	30.8	0.99	4.4	4.6	80.49

Green Grove Organics began growing and selling organic flour and grain products in 1962. Since then it has added a range of processed products such as licorice, cookies, pasta sauce, pasta, bread mixes and lamb, beef and merino wool. The company is well known on the domestic market and successfully exports a range of products to New Zealand, Norway, Singapore and Japan. The property is 1100 ha in size and has co-operated with the Australian National University in researching organic production systems. In 1998 it was presented as a world significant organic test site at the seventh International Conference of Ecology in Florence.

Lauke Mills is the last of the traditional family-owned and operated independent millers in Australia, going back to 1899. For the last 10 years it has produced a range of organic and biodynamic grain products which are widely sold on the domestic market with some export to Asia and the US. The company is currently looking at marketing several products in the UK and European markets. Its product range concentrates on wheat and rye products with a range of 'bread-making machine premixes' proving popular.

One of the largest conventional flour milling operations is also active in producing organic product for domestic and export sales. Western Milling is a major supplier of product to supermarkets and sends most of its organic product range of organic plain and self raising, wholemeal and base premixes to Japan. Its 'Tiptop' brand is well known in Australian supermarkets and for some years has carried certified organic lines.

Wholesale manufacturers and distributors

Since 1973 Spiral Foods has been well known in Australia for its focus on macrobiotics. The company has a long history of working closely with Eden Foods in the US and Muso Co in Japan. As a significant importer of organic foods, Spiral repackages and distributes across Australia to natural food, health food and supermarket customers. It also exports a range of pasta, sauces and honey to Japan, Singapore, Malaysia and Hong Kong.

Pureharvest began operations in Victoria in 1979 with a range of bulk grains, beans and nuts distributed to health food stores. It is best known as a manufacturer of organic soymilk in Australia and was responsible for financing the first biodynamic rice crop in Australia. Pureharvest manufactures a range of processed products including pasta, sauces, jams, rice crackers, corn chips and cold pressed oils, many of which are exported to the US, Asia, the Middle East, Africa, New Zealand and South America. Pureharvest is also a significant importer of organic products from Europe, the US and Japan.

Another company worthy of mention is Kadac, based in Melbourne and with a 50% share of a Sydney company, Natures Fair. As one of the largest health food distributors in Australia it turns over a huge quantity of bulk and prepackaged organic grains, flours and oils. Recently Kadac has begun production of its own range of organic products under the labels 'Lotus Organic' and 'Natures First Organic' which have been taken up by the supermarkets. Kadac is also the largest importer of organic products in Australia and distributes many of these for sale in supermarkets.

Dairy producers. The dairy industry in Australia has suffered boom and bust cycles for most of the last hundred years though the competitive global food market has taken its toll over the last decade. A recent vote by dairy farmers will see deregulation with a levy imposed per litre to pay farmers to 'exit' the industry over the next decade. Ironically this has come at a time when there is huge interest from Japan for Australia to supply organic dairy products. There is now enough interest to justify large-scale conversion of dairy farms to organic to supply the Asian markets.

Jalna and Hakea are two companies that have sold premium biodynamic and organic yogurts for the past 7 or 8 years. Their products are widely sold through specialty shops and were one of the first ranges of organic products to be consistently sold by the supermarkets 5 or 6 years ago. They produce a range of skim and mild yogurts with Hakea also featuring some very successful fruit yogurts. At this stage neither company exports though there has been some interest from Japan. Most of the organic yogurt sold in Japan is reconstituted from US organic butter and there have been expressions of interest from this market.

The fresh milk market has also been supplied with organic and biodynamic milk for a number of years by Sandhurst dairies, initially owned by the Coles Supermarket chain. Now independently owned, they still market their milk widely through the supermarkets and specialty stores. Another entrant to the market is Snowy Mountains Organic Dairy Company, which has produced a low fat (99% fat free) milk, full-cream milk and fresh cream. One problem of many organic dairy farmers is whether the dairy they deliver to consolidates organic or biodynamic milk, if not, then they have no choice but to sell the milk for the conventional market. This has frustrated many farmers and has led several to produce their own organic cheeses on farm, rather than put their milk in with the conventional.

Elgaar Farm in Tasmania produces a fine range of organic cheeses, fresh milk, yogurt, butter and cream. In addition it is an excellent grower of wheat, linseed and oats. Joe and Antonia Gretschmann, originally from Germany, settled this farm in 1986 and it has become one of the best examples of sustainable and economically effective organic farming anywhere in Australia. Tasmania boasts an excellent clean, green environment (similar to New Zealand) for producing organic foods and there are exciting plans on the drawing board to convert a large property of some 50 000 ha to organic for dairy, wine and vegetable production.

Another well-known cheese producer in Victoria since 1985 is Timboon Farmhouse Cheese. The company farms biodynamically and produces award-winning Brie and Camembert as well as fetta, torts, buetten and smokehouse cheeses. Its products appear all over Australia in gourmet delicatessens, supermarkets and specialty shops and have been used by Qantas Airlines in flight catering. It also exports to Japan, Hong Kong and Malaysia.

In South Australia the Biodynamic Farm Paris Creek company produces an excellent range of plain and flavoured biodynamic yogurts that are well recognised in specialty shops around Australia. The company is expanding its production and processing to take advantage of export opportunities into Asia. Helmut and Ulli Spranz migrated from Germany in search of a place to establish a totally biodynamic farm in 1988. A Swiss friend joined them to assist in establishing the processing facilities in 1989.

Soy products. Blue Lotus Foods in Victoria is the major producer of organic tofu and soy products, both fresh and processed. Its range, like the organic yogurts, was an early and consistent entry into the supermarkets. Beginning in 1981, and then with certified organic products in 1993, the business has outgrown its premises three times. Though the company's products are not yet exported, this may be an option for the future.

Oil. Seedex is one of the largest processed-oil companies in Australia with a healthy range of organic oils including olive, sunflower, safflower and canola. It has a successful Japanese market with safflower oil prepackaged in Australia into 500 ml plastic containers labelled and ready for sale. Another smaller company that has won numerous national small business awards is the Stoney Creek Oil Company, producing flaxseed and safflower oil, flaxseed seed meal and jojoba oil. Fred and Coral Davies began pressing their own oils with a plant on a farm and then quickly grew to be a major purchaser of organic linseed throughout Australia. They have a direct mail distribution service and now export to Hong Kong, Taiwan and Singapore.

Livestock. In Western Australia the Three Rivers Beef Pastoral Company biodynamically produces a fantastic range of quality meat. They have entered into a partnership with other like-minded meat producers to market their product to specialty food shops (not just organic) on the west and east coasts of Australia. They began marketing their product in 1996 and have already exported to Japan with inquiries from Europe and the UK, though at this stage it is too expensive to get the product such a distance. Soon they intend to produce lamb and pork through the same marketing network and are sure to become a well-known brand.

Other beef and lamb sales are well catered for in a small number of specialist butchers in the major capital cities, with the supermarkets at present planning packaging and distribution arrangements for meat. In New South Wales and Queensland the Organic Beef Export Company, trading as OBE Beef, has formed to export outback free-range organic beef to Japan and Asia. This co-operative venture involving 40 producers and over 6m ha of rangeland country is an exciting project with container-loads of beef sent to Japan. During establishment, OBE Beef received funding from an Australian government supply chain programme called Supermarket to Asia. This project is leading the way for large-scale exports of organic processed livestock from Australia and it is hoped this will lead to further investment by government in export initiatives.

Fibre. Organic wool is grown by a small number of farms who are investigating the market potential in Europe with 2000 bales per year produced thus far. An informal marketing arrangement has been developed with a number of NSW producers, one of which is Glenbye. Australia, as a major producer of wool to the world, is still cautious after stockpiling more than a million bales of wool when world prices were down. While Australia produces a considerable cotton crop there is virtually no organic cotton production to speak of. Four years ago there was production on one property but there has been no consistent follow up of this. Production of organic cotton in Australia would be welcomed, especially because of the chemical intensive nature of conventional cotton production, where there have been numerous examples of endosulphan use in cotton contaminating exported beef products. The Australian cotton crop is now 30% genetically engineered.

Breads. A range of excellent organic breads has been sold in Australia going back over the last 10–15 years. Not many bakeries have achieved certification though it is hoped this will change in the near future. John Downes started Natural Tucker, in Melbourne in 1984 and is well known for producing wood-fired sourdough organic bread. Portelli's is another bakery which started as a cottage industry and is now gearing up to supply one of the supermarket chains in New South Wales. Jack Portelli and Deb Stead started

Portelli's on their property near Bega and are supplying much of New South Wales, southern Queensland and northern Victoria specialty shops as well as Woolworths in southern New South Wales.

A favorite in central Victoria is the Himalaya Bakery located at Daylesford. Seven years ago it began production of an excellent range of breads and cakes with a European style. Himalaya mainly uses biodynamic flour and has a large following throughout specialty organic stores. The company believes in using the 'highest quality ingredients to produce the best product' and states that 'quality comes before profits'.

Well known in the supermarkets and health food stores, Country Life Bakery began focusing on wholemeal and sourdough and in the last 9 years has increasingly turned its attention to organic to satisfy customer demand. Its breads are sent interstate now and to Southeast Asia, mainly Singapore.

Pasta. There are a number of Australian brands of certified organic pasta with the most interesting being the Casalare range using native Australian plants to provide their unique flavouring. There should be export potential for these delicious and interesting pastas.

Wines. Organic wines in Australia are well known in Europe though not so well known in Australia. An organic wine tasting held in Melbourne in 1995 featured 25 organic and biodynamic wines made in Australia. Penfolds, one of the largest commercial wineries, sold some of the best organic wines at that time into Europe, including a gold medal winner. Australia now has considerable expertise in organic winemaking with healthy export sales into Europe.`

South Australia has an excellent winemaking region and two winemakers worth noting are David Bruer of Temple Bruer Wines and Leigh Verrall of Glenara Wines. Together they market their product in four states of the home market and independently export. Both companies export to the UK, many Asian countries and US. Both are certified with the Organic Vignerons Association of Australia (OVAA) and are unable to meet export demand. Tesco supermarkets in the UK sell the Temple Bruer wines at present. Both have won a number of medals at wine shows in Australia and Glenara currently exports 50% of its production. NASAA certified Captains Creek Wine is a recent example of the growing number of cool climate organic wines in central Victoria.

Herbs and tinctures. Australia also has an excellent range of organically grown culinary and medicinal herbs and teas. There is increasing interest in such products for export and many herb farms are under conversion to organic. Southern Light Herbs began in 1978 in Victoria, producing and processing a huge range of medicinal and culinary herbs. It has a network of 50 growers feeding into its processing operations and is in the early stages of negotiating with export customers in Asia. Highland Herbs in Tasmania is a newer company with a well presented product; its calendula flowers are the best ever marketed. Australians have a high regard and awareness of naturopathy in Australia with many people using medicinal herbs, hence the market demand is strong. The Pharmaceutical Plant Company has recently received certification for its range of herbal tinctures. The production of organic tinctures is an industry in itself, which has potential to bring excellent export income for Australia. The cleaner the source material, the cleaner the tincture, and with low air pollution, Australia has an advantage over many other parts of the world.

Overseas investment. Masterfoods, a subsidiary of the US Mars corporation, is investigating the production of certified organic processed sauces and dressings in Australia for sale in the US and Europe. In Ballarat in Victoria, Haku Baku is a joint Australian and Japanese noodle production company manufacturing 2000 t of organic noodles per year being shipped directly to Japan. This is one of the first Japanese noodle manufacturing plants anywhere in the world outside Japan.

Wholesalers and exporters

Organic Connection owned by Ian Diamond, who was an initial partner in Organic Wholesalers (see below), has been exporting from Australia since 1992. The enterprise exports large quantities of wheat and other grains, oils and oilseeds, fresh fruit and vegetables, nuts and dried fruits, honey, mung beans and feed grains. Recently it began production of a 1-litre tetra pack of orange juice for sale in the UK by Clearspring. The company is active in working with farmers to supply what the markets require, though the recent massive growth in organic sales in Europe has caused problems with under-supply of some wheat varieties that were oversupplied a few years ago. Organic Connection exports to the US, Asia and Europe as well as a considerable quantity of citrus to the US.

Organic Wholesalers, founded by John Williams and Ian Diamond, is now the largest fruit and vegetable wholesaler in Melbourne, supplying more than 60 specialty shops and some of the supermarket trade. They operate from the Melbourne Fresh Centre, the largest under-cover wholesale market in the Southern Hemisphere, which is popular with tourists. There is a liberal sprinkling of organic growers to be found among the conventional grower stalls, with two large wholesale stands handling produce from further afield.

Biodynamic Marketing is the second wholesale company operating from the Melbourne Fresh Centre handling Demeter biodynamic produce. This was established as a non-profit company in 1981 to distribute and promote biodynamic products and is managed by Peter Podolinsky. (The company also has a large facility at Powelltown, two hours from Melbourne, from where they distribute dried and processed Demeter products across Australia.)

In Sydney, Eco Farms is the largest distributor of fresh and processed organic and biodynamic foods with two other main competitors in Back to Eden and Marys Organics. Eco Farms also manufactures a range of organic processed foods and has recently launched an organic wheat biscuit that is selling extremely well in both Australia and New Zealand, in competition with 'Organic Vitabrits' biscuits by rival processor, Uncle Toby's. Eco Farms began producing certified processed organic products in 1996 and exports most of its products under the brand name 'Eco-Organics' to Europe, Asia, New Zealand and the US. United Organics in Brisbane, Steve's Organics in Adelaide and Bullfrog in Perth are other wholesalers of note.

13.1.9 *Domestic controls for organic products*

While exports are well regulated with 'organic export control orders' administered by the Australian Quarantine Inspection Service (AQIS), the domestic processed food and imported organic food market is unregulated. This is seen as a significant threat to the certified organic industry. One imported product, a soy drink made in Hong Kong with A$10m. sales in Australia, claims to use certified soybeans from the US, but the product itself is not certified and there is no way of knowing whether it is genuinely organic. Of more concern is the clever implication that the whole produce is certified, by claiming to use 'certified' soybeans on the cover of the package. A number of other local products have been recently launched with the same use of the word 'certified' on the packaging when in fact the whole product is not certified.

In April 1997, on behalf of the organic and biodynamic industry, AQIS submitted to the Australia and New Zealand Food Authority (ANZFA) an application to vary the Australian food standards to control the words 'organic' and 'biodynamic'. This application has unfortunately struck a technical barrier because ANZFA is a joint initiative between both countries to facilitate similar standards and better flow of food trade. ANZFA has indicated that because New Zealand does not have a national standard and no government recognition of such a standard, then it would be impossible to apply such a change to the Food Act in New Zealand.

At present the Australian organic industry through the Organic Federation of Australia is negotiating with New Zealand organic organisations to develop a plan to implement a national standard in New Zealand. There is strong support within both countries for such a domestic standard. Any such standard would almost certainly allow for the smaller producer to have an exemption if their turnover was below a certain dollar value per year. Once such a standard is in place the OFA hopes to persuade ANZFA to change the Food Standards Code to control the labelling of foods claiming to be organic and biodynamic.

Given the development of standards for the sale of organic and biodynamic foods around the world, and now internationally at Codex, there is pressure for a rapid solution to be found to this problem in both countries. One line of reasoning suggests that such a standard, and government regulation, is inevitable. Organic will continue to gain acceptance in the mainstream with consumers demanding accountability. The certified organic industry wants government involvement in application of a domestic standard, which will not cause erosion of the high standards already developed for export. Australians are only too aware of the proposal for a national organic standard in the United States, which was thoroughly rejected by the US organic industry when it was presented in draft form at the end of 1998.

The OFA has begun lobbying supermarkets as they develop their organic trade to only stock exclusively certified organic produce. At present they carry several international uncertified lines like the soymilk mentioned and more than a few uncertified organic Australian processed lines.

13.1.10 *Government policy and involvement*

While Australia was early in developing a national standard, federal government policy development has since been slow and the dollar investment in advisory and extension and research and development initiatives low. Despite the early work of AQIS, with the establishment of a national standard and the 'listed third country status' for Australia, there is, at best, token investment in organic food production systems by government. Indeed, the Australian organic industry must pay government in excess of US$100 000 per annum for it to 'oversee' the certification sector. Given the market opportunities exploding in Europe and Asia it would make good sense to commit significant R&D initiatives backed up by conversion programmes and advisory and extension services. The OFA is at present preparing a briefing paper to present to state governments around Australia arguing the case for more involvement.

At a state level the interest from government is varied, with Queensland, Western Australia and Tasmania leading the way and New South Wales, South Australia and Victoria dragging their heels. In the whole of Australia there are probably no more than ten government staff assisting organic development. Given the 10 000 or more government staff legislated to assist agriculture in this country, this small figure is most disappointing.

The relationship between the agrochemical input providers of fertilisers and pesticides and the bureaucrats and policy makers is exceptionally strong, and Australia is no exception. Due to a policy of reducing the size of government, advisory and extension services have been cut and in some cases privatised. Fertiliser and chemical companies have stepped in to the vacuum to provide this advice and spread significant misinformation against organic agriculture. It is customary for the fertiliser or pesticide agronomist to advise farmers, claiming to assist with planning their year's requirements and give free advice on nutrient and disease controls.

Organic agriculture, because it is infinitely more complex and system based, with strong connections to community principles, is difficult to quantify, research and predict. Still, research has shown the positive environmental effects of Australian organic agriculture and recent sales figures from Europe, the US and Japan have made Australian governments sit up and take notice. The OFA is targeting increased funding for organic research and development and advisory and extension services at the next federal election due in 2002.

The Rural Industries Research and Development Corporation is a federal government-owned research entity which spends A$275 000 per annum on an organic programme which funds most organic research. This is a minor amount of money considering the market opportunities and benefits to the environment, the health of farm workers and consumers.

There are no government subsidies similar to those available in Europe for organic farming. In addition the Australian government internationally has been strongly supporting the US position on global free trade with the World Trade Organisation (WTO). Certainly many in the organic industry believe that assistance during the conversion-to-organic process is an excellent way to stimulate development. The absence of government assistance programmes is stimulating the development of direct industry assistance from the retailers, processors and exporters who are desperate for supply of product. Because Australian farmers have limited regulatory controls in relation to environmental protection, some European countries have threatened to take Australia to the WTO claiming this constitutes an indirect subsidy. The Australian organic industry would welcome such action.

13.1.11 *Genetic engineering*

At present Australia has only one commercialised genetically engineered (GE) crop, Monsanto Ingard Bt cotton. More GE crops are expected in advance of regulations due to be enacted by parliament in early 2001. At present a voluntary code of practice is in place with a federal government interim regulatory office giving advice to applicants who wish to grow such crops in Australia. Secrecy is rife with the precise locations of GE trial areas a matter of contention between the Organic Federation of Australia and the government interim regulatory office. The OFA has lodged a Freedom of Information Application (FOI) to find out the trial locations and expects this to be resolved by July 2000. The reason for the secrecy is rationalised because of fear of damage to crops similar to that, which has occurred in the UK. There has been no recorded damage of sites in Australia at present. Currently there are trials of canola (Canadian oilseed rape) at 200 locations across the country. There is no way to easily assess the extent of potential contamination unless farmers conduct costly testing on all organic canola crops. It is likely that some contamination of canola crops will occur because of the high rate of pollen transfer via the wind and bees. Aventis (previously AgrEvo) and Monsanto are intending to move to a general release of canola, probably in the 2001 planting season. They have applied for approval of over 3000 ha of GE canola trials in the 2000 season and the OFA has opposed this in submissions to the government.

The OFA campaigned throughout 1999 to illustrate the importance of the organic industry to Australia, the potential of contamination from GE crops and the effect this would have on export markets. All levels of government are becoming aware of the importance of GE-free markets in Japan and Europe, and the OFA is calling for the impending regulations to comprehensively control the impact of GE crops. Government has strongly indicated its desire to control the impact on public health and safety, and the environment. The OFA and other non-government organisations want the impact on social, cultural, ethical and trade aspects to be considered. The OFA has proposed to government that a compensation fund be created, paid for by a levy on all traders in biotech foods, to cover contamination, loss of income, environment and health effects. Before the regulations are introduced in early 2001, the OFA is calling for a freeze on any further plantings of GE trials. This is justified on the basis that the interim arrangements do not take account of the effect of contamination of organic or GE-free crops, and the proposed regulations will have the capacity to do this.

13.1.12 *Future directions*

The increase in export demand and growth of consumption of organic food in Europe, the US and Asia, in particular Japan, will see investment in organic production in Australia continue slowly at first and then increase more rapidly as market demand with exports and supermarkets consolidates. Investors will continue to see opportunities to produce clean organic produce in Australia for processing and/or sale overseas with the formation of further strategic partnerships with Australian producers and processors, including the building of processing facilities in Australia. Because conversion to organic farming systems is not an overnight matter, investors will need patience and commitment.

There will be an expansion of the domestic market as the supermarkets start to seriously trade in organic product and a critical mass of organic processors gain a foothold. Longer term contracts will assist in providing opportunities for larger conversions to take place. Investors looking for export opportunities will be keen to see an expansion of the domestic market to enable them to maintain a fallback position should the export market fail for their product.

More than likely, government will find it difficult to comprehend or capitalise on the opportunities and will have to choose between the genetic engineering and organic visions for the future. Either way, organic agriculture will play an increasing role in the politics of land use and economics with serious pressure applied for policy commitment over the next

5 years. Green parties and independents may play an increasing role in backing organic farming despite Australia's system of preferential voting that makes it difficult for them to gain representation against the two major parties. Increased education of city dwellers concerning the importance of a sustainable rural sector from a sociological and environmental perspective is slowly taking place. Organic industry, consumers and environment movements have formed strong strategic partnerships and will continue to highlight the role that food production plays in caring for people's health and protecting the environment.

13.2 New Zealand

13.2.1 *Brief history of organic and biodynamic movement*

New Zealand has a well-established history of organic and biodynamic food production with the Biodynamic Farming and Gardening Association established in 1937. It also has one of the oldest organic associations in the world with the Soil and Health Association established in 1941. This organisation has been at the forefront of educating people about sustainable food production, composting and the connection between soil, food and health. The Soil and Health Association was central in the establishment of Bio-Gro, the largest certification organisation in New Zealand, which publishes, bi-monthly, New Zealand's only organic magazine where key issues surrounding the growth of the organic movement are debated. In light of the growing pressure on the term 'organic' it is undertaking the establishment of an organic domestic growers' scheme and has recently been promoting with other groups a policy vision for the whole country to go organic by 2020 (Soil and Health Association 1999).

The beginnings of organics in New Zealand are well documented by Campbell & Fairweather (1998) who claim that a number of factors influenced the expansion of organic food production in addition to the establishment of OPEG:

- The formalisation of written production standards begun in the late 1980s;
- The institutionalisation of one main labelling and inspection system (Bio-Gro);
- Organic products were already produced successfully prior to 1990 by pioneer growers for a number of key products like kiwifruit and vegetables;
- There was an existing market for organic products overseas.

Much of this early development is due to the efforts of a few. One of the more notable is Bob Crowder, who established the Biological Husbandry Unit at Lincoln College in 1977. He then became a member of the IFOAM board and 'champion for organic production (and a promoter of New Zealand organics on the world stage)' (Campbell & Fairweather 1998). It should be noted that this early development of organic agriculture in New Zealand, like in Australia, was without any subsidisation from the state (a situation that continues today in both countries). From this point on New Zealand differs from Australia in the further development of organic production systems. While Australia now has some large corporate involvement in organic processing taking place, New Zealand's corporate organic involvement began much earlier. Two large companies, Heinz-Wattie Australasia (frozen vegetables) and Zespri International (kiwifruit), took a strong interest in developing export markets with a resultant drive for conversion to organic since 1990, when the first purchase of a pea crop took place (Campbell & Fairweather 1998).

In 1995 the Organic Products Exporters Group (OPEG) was established to facilitate exports from New Zealand. Being a country that has historically exported 90% of what it produces, this was a natural step to make. OPEG is very active in New Zealand, with a progressive website and a vigorous marketing campaign at overseas trade shows. The latest organic industry development has been the formation of the Organic Federation of New Zealand (OFNZ) in late 1998, to facilitate the growth and promotion of organics in New Zealand.

New Zealanders are well focused on the future with a history of world's best practice in trading and marketing of agricultural commodities. The organic future looks set for New Zealand and it is no accident that it is outpacing Australia in organising the rapid development of organic systems to supply expanding world markets.

13.2.2 *The agricultural landscape in New Zealand*

New Zealand is a young country geologically, in complete contrast to the ancient landscape of Australia. It comprises two main islands (North and South) which, though close together, have markedly different climates. Agriculture, which concentrates on pastural practice, extends throughout both islands. Horticulture production extends from pip and subtropical fruits in the north and east through to pip and stone fruits in the south. Deep volcanic and peat soils ensure that vegetable production is also widespread.

New Zealand claims to be 'clean and green', and 'green' is well founded with lush growth from one side of the country to the other. As in Australia, 'clean' is debatable, with recent Ministry of Agriculture and Food (MAF) statistics showing that chemical use is not decreasing and, depending on how the data are interpreted, chemical use may be on the increase (Brendan Hoare 2000). However, there are positive signs that some traditional conventional agricultural systems in New Zealand are attempting to reduce their chemical usage. The kiwifruit industry is the best example, where the Kiwigreen project has resulted in consistent reductions in the use of pesticides through a quality management approach and extensive education programmes so that farmers spray only when necessary. This project was developed by Zespri International and by 1997 the total New Zealand harvest was grown under organic or Kiwigreen specifications. The use of integrated pest management such as *Bacillus thuringiensis* (known as Bt) and other oil sprays was central to the development of Kiwigreen.

During 1999, organic industry associations and environment groups led by the Soil and Health Association formed a partnership to promote New Zealand 'organic by 2020'. This initiative is a worthy vision and one well suited to the historical will of the people. New Zealanders were united in their opposition to nuclear power in the 1980s and there are strong signs of similar public mistrust in the offerings of genetic engineering.

13.2.3 *Market overview of production*

The New Zealand organic and biodynamic industry has grown rapidly since the establishment of OPEG, with the assistance of the New Zealand Trade Development Board. The value of organic production has grown from NZ$1.1m in 1990 to NZ$34m in 1997 (see Table 13.4).

Table 13.4 Value of New Zealand domestic and export market for organic production (source: Saunders *et al.* 1997).

Year	Domestic (NZ$m)	Export (NZ$m)
1990	1.0	0.1
1995	N/A	6.0
1996	N/A	10.0
1997	10.5	23.5

Organic kiwifruit production (see Table 13.5) is an excellent example of price comparisons and the increase in production of organic compared to conventional. Graeme Crawshaw, organic kiwifruit grower on the OPEG executive said in November 1999, 'The total New Zealand kiwifruit export market is worth in excess of $700 million per year and it is estimated that up to 5% will be certified organic in 2000.'

13.2.4 *Organic industry organisations*

Organic industry peak body

The Organic Federation of New Zealand (OFNZ) has been established, as the OFA in Australia, to act as the peak industry vehicle for lobbying and promotion of organics. The member organisations are the Biodynamic Farming and Gardening Association in New Zealand Inc; Bio-Gro; Soil and Health Association and OPEG. To date OFNZ has served as the vehicle for the industry to discuss the implications of various necessary decisions such as the implementation of a national standard. It exists more as an agreement for the industry to co-operate rather than a formal structure such as the OFA has in Australia with memberships and a budget. This is practical for making decisions but makes it difficult for promotion of the industry as a whole.

Certification organisations

There are three certification organisations in New Zealand. The largest, Bio-Gro New

Table 13.5 Production of trays and prices for conventional and organic kiwifruit, 1991–96 (source: Campbell *et al.* 1997).

Year	Conventional/Kiwigreen		Organic	
	No. of trays (million)	Price per tray (NZ$)	No. of trays (000)	Price per tray (NZ$)
1991	59.848	$6.08	13.069	$10.45
1992	67.272	$3.85	20.243	$7.29
1993	54.783	$4.18	51.014	$7.03
1994	55.915	$4.63	406.665	$5.88
1995	58.743	$4.22	620.095	$6.23
1996	62.437	$4.36	753.000	$7.39

Zealand, was established in 1983 and is the trading name of the New Zealand Biological Producers and Consumers Council Inc. Bio-Gro services organic production systems and its key functions include setting of standards, inspection and certification, promotion of the Bio-Gro trademark and promotion of research and development. To better service the massive growth experienced in the last 2 years, Bio-Gro recently subcontracted inspection services to SGS International. This has necessitated a steep rise in fees for licensees. In 1999 Bio-Gro gained IFOAM certification which has helped facilitate market access into Europe though there are still some issues to be resolved (see section 13.2.5).

The Biodynamic Farming and Gardening Association in New Zealand Inc was formed in 1937 and incorporated in 1945 to further the biodynamic method of agriculture, horticulture and forestry as elucidated by Rudolf Steiner. It is responsible for the Demeter logo in New Zealand and, while they have a membership of over 800, their licensees number only 37. Many of their members are small producers dedicated to the concept of healthy food for the local community. There is no incentive or requirement at this stage to get certified. A member of the association executive sits on the board of Bio-Gro and the OPEG executive, showing a healthy level of co-operation. Soil and Health, the founder of Bio-Gro, also maintain a seat on Bio-Gro.

The third certification organisation, AgriQuality, is a MAF-owned enterprise set up in 1998. Its standards are based on the Codex guidelines and it is able to certify to importing country requirements. This development is seen with varying degrees of suspicion and hostility from other organic industry sectors, largely because government has provided very little assistance to the established organic and biodynamic industry bodies in New Zealand to gain access to European markets. When asked to provide accreditation for the existing certification bodies, the New Zealand government has consistently said such accreditation can only come at a huge price. At the same time an arm of government (AgriQuality) was developed, effectively competing with the existing established certification bodies. AgriQuality had 22 licensees at the end of 1999.

In summary, the number of certified licensees has grown threefold since the early 1990s (see Table 13.6).

13.2.5 *Regulations for production and processing*

At present there are no regulations for production and processing of organic foods in New

Table 13.6   Total number of licensees in New Zealand.

	1992	1994	1999
Bio-Gro	232*	253*	748†
Biodynamic	N/A	N/A	37†
AgriQuality	N/A	N/A	22†

*Campbell & Fairweather (1998).
†Direct advice, current as at November 1999 from each of the certification organisations.
NB There are no figures available for the area of land under certification in New Zealand.

Zealand. As mentioned in section 13.1.5, discussions with the New Zealand and Australian organic industry, begun in 1999, are continuing through 2000 and it is hoped that a national standard will be implemented in the next 12 months. The initial prime motivator for such a standard from a New Zealand perspective is to maintain market access into European and soon Japanese markets. This has now grown to include protecting of the term 'organic'. Japan has indicated that it will require evidence of equivalence to their standards by April 2000.

There is a history of problems associated with a small but significant percentage of New Zealand exports to Europe that have not met entry requirements or have required documentation that was not correctly supplied from New Zealand. On more than one occasion New Zealanders have had to board a plane to Europe to expedite entry of New Zealand product sitting on the docks.

In order to meet equivalence requirements, Bio-Gro applied for IFOAM accreditation, which it received in early 1999. Unfortunately EU regulations mean that there is no certainty that IFOAM accreditation will guarantee automatic equivalence status. The Demeter product with no third party auditing has no means of demonstrating equivalence. AgriQuality makes the claim that it can give equivalence certification to meet importing country requirements but this has not yet been put to the test. OPEG's executive director, Jon Manhire, said in a September 1999 press release, 'OPEG is aware that some exporters are adopting a conservative approach to European Union markets because of regulatory obstacles. In the past financial year … ensuring access to the EU absorbed a considerable amount of the group's scarce resources, and this effort is continuing.' As a result of the stricter requirements governing ongoing access to the European and Japanese markets, the New Zealand organic industry will almost certainly be required to adopt a national standard or gain further equivalence accreditation externally.

13.2.6 *Market overview of exports*

New Zealand is focused on exports with domestic organic consumption accounting for only 15–20% of production. Without a successful organic export market New Zealand could not afford to convert to organic. New Zealand growth rates are similar to other parts of the world with NZ$65m predicted for exports alone by 2000/01 (see Fig. 13.2). Jon Manhire stated in September 1999, 'Last year's result met OPEG's projections, but could have been greater. This is frustrating, because there is a global shortfall in supplies of organic food, at a time when demand is surging and premiums over non-organic product are still considerable – anywhere from 20% to 100%.' OPEG chairman Stuart Abbott went on to say, 'There are signs that the second phase of the New Zealand organics industry is emerging. It will come from companies that have been watching international trends and can no longer afford to be without an organic option in their product portfolio.'

The major percentage of exports goes to Japan and Europe, with Australia and the US around 5% and 4% respectively (see Fig. 13.3). The reason for the high Japanese percentage is predominately due to the export of organic frozen vegetables by Heinz-Wattie Australasia, and organic kiwifruit.

In Fig. 13.4 the breakdown of organic exports is by product; note the large processed vegetables percentage compared to fresh, and the huge percentage of fresh fruit. In recent years the export of fresh fruit to Europe, particularly apples and kiwifruit, has surged.

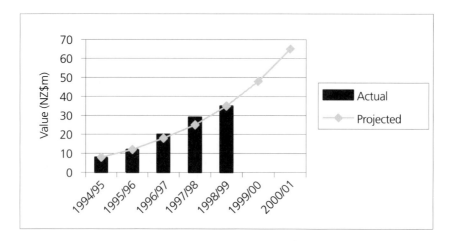

Fig. 13.2 OPEG Annual Survey 1999 (source: OPEG member survey 1999).

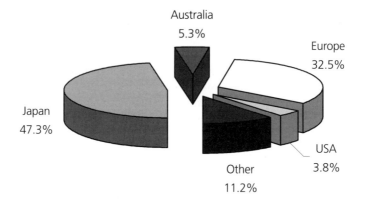

Fig. 13.3 New Zealand organic exports by market (source: OPEG member survey 1999).

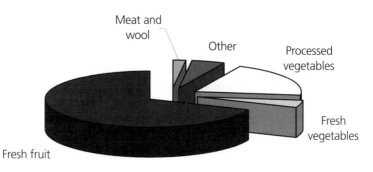

Fig. 13.4 New Zealand organic exports by product (source: OPEG member survey 1999).

13.2.7 *Market overview domestic consumption*

The only figures for domestic consumption come from those quoted in Table 13.4 by Saunders *et al.* (1997). Brendan Hoare, president of Soil and Health, reports recent conversations with the domestic retail industry which suggest the current domestic market has grown to around $100m. In New Zealand, as in Australia, most retailers, growers and wholesalers have reported significant increases in the sales of organic foods. Research is required to verify and quantify the contribution the domestic market makes to overall demand for organic products. Organic retailer and farmer, Jim Kebbel, began his involvement in the organic industry 10 years ago as a farmer, but frustration with the volatile nature of the market prompted him to start his own store in Wellington in the 1980s. His business (Commonsense Organics) has continued to grow and is now in new premises trading as a medium-sized organic supermarket. Home delivery is an important contributor to the growth of his business. In addition to Wellington, all the other major centres feature busy organic food stores and some supermarkets, which are now beginning to stock organic. A number of years ago when Heinz-Wattie Australasia began production of their organic frozen vegetables range, supermarkets in both New Zealand and Australia stocked them with very little success. The domestic consumer in both countries at that time was not ready for frozen organic food in supermarkets. As organic becomes acceptable to the mainstream, sales of frozen and other highly processed convenience-style organic products are expected to increase.

There will always be a concerned minority of organic consumers worldwide who will continue to demand fresh, locally produced organic product and would not dream of buying frozen imported product. These people will continue to raise global awareness of the issues of energy use in food production and transport by supporting the concept of organic as being much more than minimal chemical use. They see organic as a way of life encompassing such principles as fair trade, social responsibility, minimal processing and transportation, consumption near production, and so on.

13.2.8 *New Zealand organic industry profiles*

Processors

There is an excellent range of processed products available in New Zealand, with some exceptions similar to Australia, such as processed breakfast cereals. While New Zealand relies upon Australia for most of its organic grain supplies, it has developed an innovative approach to processing. One example is the Only Organic baby food range. This company, established only 5 years ago, now exports a range of 20 baby foods to Australia and Asia. While originally planned to focus on the UK market, the overvaluing of the NZ dollar put this market out of reach, leaving Australia as the focus. Heinz in Australia is about to release its own certified organic baby food that will compete in supermarkets with the Only Organic range. Only Organic now intends to develop products for older palates.

The earlier-mentioned Heinz-Wattie Australasia Ltd developed its first organic products for export in 1991 and has moved on to command a considerable presence in Japan worth tens of millions of dollars. Heinz-Wattie Australasia, as the name suggests, operates in New Zealand and Australia, though to date the processed frozen organic export products are sourced from New Zealand. The company now has 50 growers producing certified organic product on more than 2500 ha of land. The principal crops are peas, carrots and sweetcorn, with more on the way. The company has developed the 'Grow Organic with Wattie's' programme, resulting in research to develop weed, pest, disease and fertility methods and providing technical assistance with certification to Bio-Gro standards. Premiums offered are 20–100% depending on the crop.

Wholesalers and exporters

Probably the best-known exporter of organic products in New Zealand is Zespri International. In 1991 Zespri International (known then as the Kiwifruit Marketing Board) began a trial marketing programme with 13 000 trays. Production in 2000 is expected to be more than 2.5 million fully certified organic trays with the average premium pay-out over the years equating to 67%. The largest market is Japan, developed since 1994. Stuart Abbott, project manager for Zespri, believes the greatest threat to markets is product integrity and cites the example of Chile where product exported as organic was detected to contain chemical residues. There was a subsequent loss of trust, ending exports of Chilean organic fruit to Japan. One of the benefits of organic kiwifruit production is that fruit losses are half that of conventional production systems, which can be attributed to more wholesome orchard inputs and the absence of chemically soluble fertilisers (Country News 1999).

While kiwifruit may be the largest export crop, another Auckland based company is developing export markets with other fruits into Europe at an extraordinary rate. Freshco employs 13 people and has a sales office in Europe, with its New Zealand Organics Ltd division handling organic sales. In 1999 the sales figure exceeded NZ$12m, continuing the trend of growth at 150% per year for the last 3 years. One of its European customers is Tesco, and while one-third of its turnover is presently organic, the intention is to be exporting 100% organic and/or environmentally certified product within 5 years. Freshco exports to 12 countries including the US, Canada, Japan, UK, France, Germany and other Pacific Rim and EU countries (Country News 1999).

As in Australia, on-farm processing is popular, with CoralTree Organic Products Ltd being an innovative example. They are apple orchardists now producing an excellent apple cider vinegar, apple juices (including sparkling) and pickled products such as onions, gerkhins and beetroot. The company has begun exporting to Australia and would like to expand its markets further. All over New Zealand are similar examples of small processed organic on-farm enterprises, many of which are capable of exporting.

Dairy. An excellent range of organic dairy products is available from butter through to cheeses, milk and yogurt. Biofarm is the largest dairy processor established in New Zealand with a 400-acre farm situated in the Manawatu region, renowned for its rich soils and temperate climate. The farm was established in 1977 and after 5 years the enterprise began using biodynamic principles. In 1986 it gained Bio-Gro certification and entered the market with its Biofarm Acidophilus Yogurt. In the last 2 years the company enjoyed a 60% growth in sales. Production is in excess of 1m litres per year with all processed on farm into yogurts and pasteurised whole milk.

Livestock. New Zealand has traditionally provided sheep and dairy products for the export market so it is not surprising to see well-developed organic meat products available throughout New Zealand. Harmony is a company founded in 1996 by two Dutch immigrants, Ben and Anna van Toledo, who have established the enterprise as a wholesale only, organic meat processing company. Their range of fully traceable pork, lamb, beef and poultry products are supplied to small retail outlets through to supermarkets.

There is also recognition of the potential for organic fibre products, with Treliske based at central Otago. This 3000-acre family farm runs Merino, Romney and coloured sheep flocks with Angus and Hereford cattle. Fleeces are then selected by hand and spun into undyed, unbleached quality yarns which are hand and machine knitted and exported mainly to the US and Japan.

Honey. Honey is an abundant product in New Zealand with more than 300 t produced each year organically. In addition, Manuka honey has won worldwide acclaim for its antibiotic and wound healing properties. Waitaki Apiaries produce half of the organic honey crop from 3000 hives in wilderness areas in the remote valleys of the plateau country surrounding the majestic Southern Alps. The honey is exported in bulk to Japan and Denmark in two flavours, White Clover and Vipers Bugloss. From Denmark it is then sold throughout Europe.

Wine. There are excellent organic wines produced in New Zealand, with many exported. One winery with many awards to its credit is Richmond Plains, owned by the Holmes Brothers in Nelson province in the South Island. The winegrowers of the region show remarkable co-operation by sharing processing facilities. Their first vintage was in 1995 and they now export to Australia, Japan and the UK. The Richmond Plains collection includes three whites – Sauvignon Blanc, Autumn Harvest and a barrique-fermented Chardonnay, and two reds – Pinot Noir and a Bordeaux style blend named 'Escapade'.

Beer. At Nelson in the South Island the first certified organic beer produced in the southern hemisphere has been produced. At Founders historic park, Founders Brewery has produced a hand crafted beer with two unique flavours, Tall Blonde and Long Black, for export to Austria. Another flavour, Redhead, is available on tap at the brewery.

Bread. The only certified bakery is Paraoa Bakehouse established in 1996 in Wellington. It produces a delicious range of breads using filtered water, marketed throughout New Zealand under the brand name Purebread. Their product range includes traditional kibble wheat and rye breads, oatmeal and honey, rye and rice, flat breads, fruit breads, granola and tasty fruit and nut bars.

Herbs. New Zealand has an emerging organic herb market and Coralie is leading the way with a range of certified lavender and rosemary essential oils and culinary herbs produced locally. Processing takes place on farm and all products are guaranteed 100% New Zealand grown.

13.2.9 *Domestic controls for organic*

There are no controls on production or labelling of organic foods in New Zealand, however the introduction of a national standard would facilitate the introduction of domestic standards in both New Zealand and Australia. One hurdle to overcome is the considerable variation of opinion about the means of implementation and benefits of such a national standard, and a strong desire for the status quo to minimise 'state' involvement. It should be said that, as soon as organic and biodynamic sales constitute a significant market share in any country, it will be absolutely necessary for controls to be introduced. Europe and soon the US, Canada and Japan are all introducing domestic controls, leaving Australia and New Zealand behind. Consumers are now the main drivers of the organic industry and they will continue, quite rightly, to demand accountability and certainty that the products they buy are genuine.

13.2.10 *Government policy and involvement*

The involvement of the government in developing organic production in New Zealand has been limited. The New Zealand Trade Development Board was involved in establishing OPEG. In the last couple of years there are signs that other departments such as MAF are interested in exploring options to further develop organic agriculture. This conservative approach on behalf of MAF is strikingly similar to that taken in Australia, but may be about to change.

   The New Zealand voting system, unlike the Australian, favours smaller political parties, with approximately half the parliament elected via an electorate candidate vote and the other half via a party vote. This has resulted in the Green Party holding 7 seats out of the 120-seat parliament at the last election in late 1999. The Labour Party is governing with a minority partnership with the Alliance Party, and the Greens hold the balance of power. This is in total contrast to Australia, which has seen two major, increasingly similar parties dominate, with the current national government being conservative. Two of the Green parliamentarians are organic farmers and the Alliance Party's Consumer Affairs Minister is said to have been the founder of an organic co-operative during student days. A significant change in government policy can be expected shortly, with organic and environment groups lobbying for New Zealand to be organic by 2020.

13.2.11 *Genetic engineering*

From a distance, if you were reading reports from the various government departments involved in biotechnology research, you could be excused for thinking New Zealand was embracing genetic engineering with great enthusiasm. Yet, despite a strong policy of support from the previous government and business, New Zealand has developed more slowly on genetic engineering than Australia, with far fewer field trials undertaken. There is no equivalent crop to the Ingard Bt Cotton grown and the thousands of hectares of GE canola crop trials in Australia. Instead there has been a lot of interest in New Zealand in genetically engineering dairy cattle and fish. Fortunately for New Zealand, such research has not reached the stage that there is a danger of contamination which could threaten the ability of the country to market itself as GE-free.

   There is keen consumer resistance to genetic engineering, with 5000 submissions from a population of 3.5 million New Zealanders made to the Australia and New Zealand Food Authority asking for full labelling (compared with 1000 Australians out of 18 million). Genetic engineering was made an election issue in November 1999, and the Greens have presented a petition to parliament with more than 92 000 signatures calling for a royal commission to be established to 'enquire into and advise on the ethics, scientific uncertainties, health risks and benefits, environmental effects, and economic repercussions of genetic engineering of food crops, animals, and other organisms'. It calls for the commission to 'hold public hearings in the main centres in New Zealand with cross-examination of evidence'. Until the royal commission has reported there should 'be a moratorium on the release or field trials of transgenic crops, animals or other organisms and on the approval of any further transgenic foods for sale in New Zealand'.

13.2.12 *Future directions*

New Zealand is well positioned to take advantage of the world market for organic foods, with an excellent image internationally and a focused industry expansion plan. New Zealand will aggressively market its organic products internationally, which will be necessary as organic becomes a globalised product. Already there are signs of vigorous competition between countries, and New Zealand is positioned to cope well, with the OPEG group established and performing.

The environment movement, already seamlessly integrated with organic agriculture, will be effective in influencing government policy as witnessed by the recent election that left the Green Party holding the balance of power. The Labour–Alliance government will begin to actively support organic agriculture by applying regulatory protection to organic production systems that may be contaminated by genetically engineered crops and overspray.

The corporate food industry that saw the organic opportunity in the late 1980s will continue to invest in conversion and supply management strategies aiming to export to Europe, the US, Japan and other parts of Asia including Australia. There will be an increase in strategic long-term partnerships with overseas investors and organic processors or wholesalers to buy direct. This will ensure a steady increase in confidence and development of the organic industry and see the vision of New Zealand totally organic by 2020 moving rapidly from fantasy to reality.

References

Alexandra, J. (1999) *Environmental Management Systems For Australian Agriculture – Issues and Opportunities.* Briefing paper for the Environmental Management Systems in Agriculture – Current Issues: Future Directions Workshop Ballina, Australia, May 1999.

Baker, K. (2000) Proprietor of Organically New Zealand, Levin, New Zealand. Personal communication.

Campbell, H. & Fairweather, J. (1998) *The Development of Organic Horticultural Exports in New Zealand.* Research Report No. 238, Agribusiness & Economics Research Unit, Lincoln University, Canterbury, New Zealand.

Campbell, H., Fairweather, J. & Steven, D. (1997) *Recent developments in organic food production in New Zealand: Part 2, Kiwifruit in the Bay of Plenty.* Studies in Rural Sustainability – Research Report No. 2, Department of Anthropology, Otago University, New Zealand.

Country News, 11 November 1999. Organic Expo '99 liftout, Bay of Plenty, New Zealand.

Hassell and Associates (1996) *The Domestic Market for Australian Organic Produce, An Update.* A Report for Rural Industries Research and Development Corporation, Australia. RIRDC Research Paper No. 96/1.

Hoare, B. (2000) Personal communication. President of Soil & Health Association, New Zealand.

Macarthur Agribusiness and Quarantine and Inspection Resources Pty Ltd (1999) *Organic Certifiers – AQIS Charges Review, Sub-Program 2.5 Organic Produce.* A Report for Rural Industries Research and Development Corporation, Australia. RIRDC Project No. MS990–20.

McCoy, S. & Parlevliet, G. (1999) *The Export Market Potential for Clean and Organic Agricultural Products.*

A Report for Agriculture Western Australia and Rural Industries Research and Development Corporation, Australia. RIRDC Project No. DAW 85A.

Melbourne Age (1999) Dr Graham Harris, Head of Land & Water Research, Commonwealth Scientific Industrial Research Organisation (CSIRO), reported in the *Melbourne Age* 8 August 1999.

OPEG Member Survey 1999. Produced by the Organic Products Exporters Group, New Zealand. http://www.organicsnewzealand.org.nz

Saunders, C., Manhire, J., Campbell, H. & Fairweather, J. (1997) *Organic Farming in New Zealand: An evaluation of the current and future prospects including an assessment of research needs.* MAF Policy Technical Paper 97/13. MAF, Wellington, New Zealand.

Soil & Health Association (1999) *Soil & Health*, Nov/Dec 1999, Vol 58, No 6. (Journal of the Soil & Health Association of New Zealand.)

Williams, J. (1999) Personal communication. Proprietor of Organic Wholesalers, Melbourne, Australia.

Contacts

Organic websites

Organic certification organisations

International Federation of Organic Agricultural Movements (IFOAM)	www.ifoam.org
Organic Food Federation	www.organicfood.co.uk/off/index.html
SKAL	www.com/english_skal/index.htm
Soil Association	www.soilassociation.org

Organic delivery companies

Abel & Cole	www.abel-cole.co.uk
Fresh Food Company	www.freshfood.co.uk
Organics Direct	www.organicsdirect.co.uk
Simply Organic	www.simplyorganic.net
The Organic Shop	www.theorganicshop.co.uk

Organic food and drink producers

Alara Muesli	www.alara.co.uk
Cafedirect	www.cafedirect.co.uk/organic/index.html
Cascadian Farm	www.cfarm.com
Clipper Teas	www.clipper-teas.com
Enjoy Organic	www.enjoyorganic.com
Finest Organics	www.finestorganics.co.uk
Graig Farm	www.graigfarm.co.uk
Green & Black's	www.earthfoods.co.uk/gbs.home
Goodness Foods	www.goodness.co.uk
Juniper Green	www.junipergreen.org
Nature's Path	www.naturespath.com
MD Foods	www.harmonie.co.uk
Muir Glen	www.muirglen.com
Organix Brands	www.babyorganix.co.uk
Organic Farm Foods	www.organicfarmfoods.co.uk
Pure Organics	www.pureorganics.co.uk
Rachels Dairy	www.rachelsdairy.co.uk
Seeds of Change	www.seedsofchange.co.uk
The Village Bakery	www.village-bakery.com

The Stamp Collection	www.stamp-collection.co.uk
UK5	www.uk5.org
Urtekram	www.urtekram.dk
Whole Earth Foods	www.earthfoods.co.uk

Organic ingredient suppliers

Beta Pure	www.betapure.com
Biofood Net	www.biofood.net/english/bus.htm
Community	www.communityfoods.co.uk
Green Trade Net	www.green-tradenet.de
Organic Trade	www.organictrade.co.uk
Trade Organex	www.tradeorganex.com
Tradin	www.tradinorganic.com

Organic media

Organic Food	www.organicfood.co.uk
Organic Living	www.organicliving.co.uk
Natural Foods Merchandiser	www.nfm-online.com
Natural Products News	www.naturalproducts.co.uk
Simply Food	www.simplyfood.co.uk/organic

Miscellaneous

Canadian Organic Advisory Board	www.coab.ca
Consumers International	www.consumersinternational.org
Consumers Union	www.consunion.org/food
Ecoweb	www.ecoweb.dk/english
Fairtrade Foundation	www.gn.apc.org/fairtrade
Green Books	www.greenbooks.co.uk/organic/htm
Henry Doubleday Research Association (HDRA)	www.hdra.org.uk
Organic Consumers Association	www.organicconsumers.org
Organic Trade Association	www.ota.com
Organic-UK	www.organic.mcmail.com
The Organic Consultancy	www.organic-consultancy.com
True Food	www.truefood.org

Retailers

As Nature Intended	www.asnatureintended.ltd.uk
FDB	www.fdb.dk/natura
Sainsbury's	www.tasteforlife.co.uk/organics
Waitrose	www.waitrose.co.uk/new/organics/index.htm
Wild Oats	www.wildoats.com

Trade shows and events

Healthy Ingredients Europe	www.fi-events.com/hi
Organic Business 2000	www.fi-events.com/organic
Organic Food and Wine Festival	www.organicfoodwinefestival.co.uk

Universities with organic departments

Aberystwyth University	www.wirs.aber.ac.uk/research/organic.shtml
Scottish Agricultural College (SAC)	www.sac.ac.uk/cropsci/external/organic/ default.htm

UK approved sector bodies

United Kingdom Register of Organic
Food Standards (UKROFS)
c/o MAFF
Room 114, Nobel House
17 Smith Square
London SW1P 3JR
Tel: 0207 238 6004
Fax: 0207 238 6148

Scottish Organic Producers Association
Milton of Cambus Farm
Doune
Perthshire
FK16 6HG
Tel: 01786 841657
Fax: 01786 842264

Soil Association Certification Ltd
Bristol House
40–56 Victoria Street
Bristol
BS1 6BY
Tel: 01179 142400
Fax: 01179 252504

Organic Farmers and Growers Ltd
50 High Street
Soham
Ely
Cambridgeshire CB7 5HP
Tel: 01353 720250
Fax: 01353 720289

Organic Food Federation
Unit 1
Manor Enterprise Centre
Mowles Manor
Etling Green
Dereham
Norfolk NR20 3EZ
Tel: 01362 637314
Fax: 01362 637398

Bio-Dynamic Agricultural Association
The Painswick Inn Project
Gloucester Street
Stroud
GL5 1QG
Tel: 01453 759501
Fax: 01453 759501

*Irish Organic Farmers and Growers
Association*
Harbour Building
Harbour Road
Kilbeggan
County Westmeath
Ireland
Tel: 00 353 506 32563
Fax: 00 353 506 32063

Organic Trust Limited
Vernon House
2 Vernon Avenue
Clontarf
Dublin 3
Tel: 00 353 185 30271
Fax: 00 353 185 30271

*Food Certification (Scotland) Ltd
(Organic Certification of Farmed Salmon in
the UK)*
Redwood
19 Culduthel Road
Inverness
IV2 4AA
Tel: 01463 222251
Fax   01463 711408

Index